Introduction to Waves: Sound, Light, and Beyond

INTRODUCTION TO WAVES: SOUND, LIGHT, AND BEYOND

Dr. SachchidaNand Shukla

ARCLER PRESS

www.arclerpress.com

Introduction to Waves: Sound, Light, and Beyond

Dr. SachchidaNand Shukla

Arcler Press
224 Shoreacres Road
Burlington, ON L7L 2H2
Canada
www.arclerpress.com
Email: orders@arclereducation.com

ISBN: 978-1-77469-936-2 (Hardcover)

ABOUT THE AUTHOR

Dr. SachchidaNand Shukla, is presently working as Professor, Department of Physics & Electronics, Dr. Ram-Manohar Lohia Avadh University, Ayodhya, UP, India. He did his Masters in Physics (Electronics) in 1988 and Ph.D. in 1992 from the same university. Dr. Shukla holds 27 years' experience of teaching M.Sc. (Physics) and M.Sc. (Electronics) students and 3 years experience of teaching B.Tech., MCA and B.Sc. (Electronics) students. He has published 85 research papers in peer-reviewed/ indexed journals of International/National repute & conference proceedings and 02 books. In his supervision 14 research scholars have been awarded Ph.D. degree. In addition to it he is the recipient of Best Scientist National Award (2018) of IRDP Group of Journals, Chennai and Maatee Ratan Samman (2017). He has also been selected as Fellow of IACSIT (International Association of Computer Science and Information Technology, Singapore) and Associate Fellow of IAPS (International Academy of Physical Sciences) in 2018. In view of Dr. Shukla's academic achievements his employer institution, Dr. RamManohar Lohia Avadh University, has conferred upon him the 'Certificate of Appreciation' in 2018. Besides having a wide exposure to various key positions of University administration like Pro Vice Chancellor, Registrar, Director College Development Council (CDC), Coordinator UGC and RUSA, Head of Physics Department etc, Dr. Shukla has membership of 08 academic bodies of international repute. To name a few are ISCA (The Indian Science Congress Association, Kolkata, India), IETE (The Institution of Electronics and Telecommunication Engineers, New Delhi, India), NASI (The National Academy of Sciences, India, Allahabad, India), IAPS (International Academy of Physical Sciences, Allahabad, India and SCIEI (Science and Engineering Institute, Hong Kong, SAR of China).

TABLE OF CONTENTS

LIST OF FIGURES

LIST OF TABLES

LIST OF ABBREVIATIONS

5G	Fifth-Generation
7B	Seventh-Grade Science Curriculum
AI	Artificial Intelligence
CCD	Charged Coupled Devices
CD	Compact Discs
CT	Computed Tomography
DNA	Deoxyribonucleic Acid
DVD	Digital Video Devices
EM	Electromagnetic
ESWT	Extracorporeal Shockwave Therapy
EUSIPCO	European Signal Processing Conference
FCC	Federal Communications Commission
GR	General Relativity
HIFU	High-Intensity Focused Ultrasound
HPC	Harmonic Polynomial Cell
IR	Infrared
ITU	International Telecommunication Union
JWST	James Webb Space Telescope
LIGO	Laser Interferometer Gravitational-Wave Observatory
NDT	Non-Destructive Testing
OTEC	Ocean Thermal Energy Conversion
PET	Positron Emission Tomography
PV	Photovoltaic
QKD	Quantum Key Distribution
SAR	Synthetic Aperture Radar
SERCA	Sarco/Endoplasmic Reticulum Ca2+
SKA	Square Kilometer Array
STEM	Science, Technology, Engineering, and Mathematics

TMS	Transcranial Magnetic Stimulation
UCMMT	UK-Europe-China Workshop on Millimeter Waves and Terahertz Technologies
UV	Ultraviolet
WCNC	Wireless Communications and Networking Conference
WDM	Wavelength Division Multiplexing
WFIRST	Wide Field Infrared Survey Telescope
XRF	X-Ray Fluorescence

PREFACE

The book "Introduction to Waves: Sound, Light, and Beyond" is an exploration into the fascinating world of wave phenomena that surround and interact with us every day. This subject delves into the nature and characteristics of different types of waves, including sound waves, which vibrate air molecules to facilitate auditory communication, and light waves, which illuminate our world and provide the basis for visual experiences. Beyond these familiar types, the study extends to more complex phenomena like electromagnetic waves that encompass microwaves, radio waves, X-rays, and more, responsible for much of our modern communication and diagnostic technology. This field combines aspects of physics, mathematics, and engineering, providing insight into how waves transfer energy and information, affecting everything from the creation of music to the functionality of our cell phones. Whether it's understanding the basic wave properties like amplitude, wavelength, and frequency, or grasping the nuances of wave interference and polarization, studying waves paves the way for a deeper understanding of the universe and our technological advances.

This book is a thoughtfully composed eight-chapter journey into the intriguing world of wave phenomena. Each chapter unfolds a new layer of understanding, imparting knowledge from basic principles to advanced applications and future prospects.

Chapter 1 lays the groundwork, familiarizing the reader with the foundational concept of waves. Chapter 2 takes the reader on a fascinating exploration of wave phenomena in nature, highlighting how waves manifest in our everyday surroundings.

Chapter 3 dives into the fascinating world of sound waves, dissecting their generation, propagation, and impact on our auditory experience. Chapter 4 turns its focus towards light waves, illuminating their role in the formation of visual stimuli and the intricate process by which we perceive the world around us.

In Chapter 5, readers will be drawn into the peculiarities of wave behavior, particularly the phenomena of interference, refraction, and diffraction. Chapter 6 pushes beyond sound and light to unveil the existence and effects of other types of waves such as electromagnetic and gravitational waves, further broadening the wave spectrum.

Chapter 7 details the intricate connection between wave phenomena and modern technological applications, demonstrating how understanding waves has powered advancements in telecommunications, medical diagnostics, and numerous other fields. Finally, Chapter 8 looks towards the horizon, extrapolating current knowledge to anticipate future developments and challenges in the study of waves.

Written by experts in the field, "Introduction to Waves: Sound, Light, and Beyond" offers a comprehensive, detailed, and engaging coverage of the subject. It combines rigorous research with practical insights to deliver an outstanding resource for students, educators, researchers, and anyone intrigued by the omnipresence and influence of wave phenomena in our lives. This book aims to inspire curiosity, stimulate discussion, and foster a deeper understanding of the world around us through the lens of wave science.

CHAPTER 1

Introduction to Waves

Contents

1.1 INTRODUCTION

Waves can be described as perturbations that propagate energy through a given medium or space while not inducing any significant displacement of the medium itself. Mechanical and electromagnetic waves are two distinct categories that can be used to classify waves. Mechanical waves necessitate a physical medium to propagate, while electromagnetic waves can travel through a vacuum. Waves possess distinct attributes, including amplitude, wavelength, frequency, and speed. Amplitude refers to the maximum displacement of a wave from its equilibrium position. Wavelength represents the spatial interval between successive points within a wave. Frequency denotes the number of cycles traversing a specific point within a given time frame. Lastly, speed is determined by the product of frequency and wavelength (Parker, 2009).

Waves demonstrate a range of phenomena, such as reflection, refraction, diffraction, and interference. Reflection refers to the bouncing back of waves upon encountering a boundary. Refraction involves the alteration of both direction and speed as waves pass through different media. Diffraction pertains to the bending of waves around obstacles or apertures. Interference encompasses the combination of waves, resulting in either constructive or destructive effects. These characteristics enable the utilization of waves in various fields (Crampin, 1984).

Electromagnetic waves are employed in communication to facilitate wireless transmission, encompassing various applications such as radio broadcasting, television broadcasting, cellular telephony, and Wi-Fi connectivity. Medical imaging techniques utilize various types of waves, such as X-rays and ultrasounds, to facilitate the visualization of internal anatomical structures. Sonar technology utilizes acoustic waves to facilitate underwater navigation, cartography, and the identification of submerged entities. Seismic waves play a vital role in the field of seismology as they are instrumental in the examination of the Earth's internal structure and the identification of seismic events (Wu & Aki, 1988).

A comprehensive comprehension of waves is imperative for progressing scientific knowledge and advancing technologies that exploit wave phenomena to enhance diverse facets of human existence. Waves, ranging from basic undulations on liquid surfaces to intricate electromagnetic transmissions, profoundly influence the structure of the natural environment and our engagement with it (Knobel, 2000).

1.2 BASIC WAVE CONCEPTS

The initial segment of this unit is associated with two significant objectives. There are two primary objectives in this study. The first objective is to comprehensively understand wave phenomena and the methods employed in their analysis. The second objective is to comprehend the mathematical depiction of one-dimensional harmonic waves proficiently. The intention is to employ this mathematical model as a fundamental instrument for comprehending the principles underlying the phenomena of light, sound, and various other wave propagation forms throughout the course (Tsurutani & Lakhina, 1997).

A material wave is a prevalent form of internal motion a substance exhibits. The medium refers to the substance through which the wave propagates. For the existence of material waves, there must exist forces acting between adjacent particles within the medium. To facilitate comprehension, we will analyze the propagation of a disturbance and its consequent manifestation as wave-like phenomena. This will be achieved by assigning distinct colors to three distinct portions of a medium, denoted as sections 1, 2, and 3 in Figure 1.1 provided (Keyrouz & Caratelli, 2016).

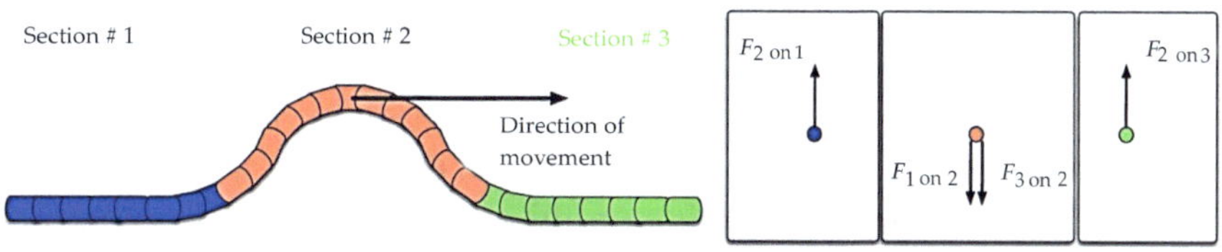

Figure 1.1. Representation of a material wave (Source: The Libre Texts libraries, Creative Commons License).

The state of equilibrium is achieved by the medium when it remains undisturbed. If the medium mentioned above is considered a section of a rope, it would be in a state of equilibrium and lie flat. The disturbance of this medium can be initiated by applying a shaking motion to one end of the rope. In the depicted wave instant, as shown in Figure 1.1, Section 2 experiences a displacement from its equilibrium position due to its interaction with Sections 1 and 3. The shift of the medium as the waves pass through it is commonly referred to as the distance from equilibrium (Comtois et al., 2005). According to Newton's third law, the downward force exerted on Section 2 results from the forces exerted by Sections 1 and 3. The force diagrams above demonstrate the presence of opposite arrows, indicating that F1 on 2 equals the negative of F2 on 1, and F3 on 2 equals the negative of F2 on 3. Therefore, it can be inferred that Section 2 will experience a downward acceleration, ultimately returning to equilibrium. As a result of this phenomenon, it can be observed that Section 3 experiences an upward acceleration, leading to its subsequent displacement resembling that of Section 2 after a certain time. Thus, the disturbance propagation from Section 2 to Section 3 occurred without the concomitant displacement of individual components of the medium (Thoads & Roedel, 1999).

One might question the underlying reasons for the interplay of forces between the sections. The genesis of this force can be attributed to the pair-wise interactions elucidated in the context of 7A. Atoms possess a distinct equilibrium separation distance, which they prefer, and actively resist any external forces attempting to alter this distance. Elongating or condensing the medium induces the atoms to apply forces on adjacent atoms and oppose external forces, commonly called restoring forces. Atoms have been grouped into three sections for convenience, but the same discussion can be applied to individual atoms. The restoring forces that govern wave phenomena are generally more pronounced in solid substances but insignificant in most gaseous substances (Pedlosky, 2003).

The concepts of material waves that we introduce resemble those previously introduced in the context of motion, forces, and atoms. The significant emphasis on waves as an independent phenomenon stems from the inherent simplicity of analyzing waveforms instead of trying to illustrate force diagrams for numerous sections within a given medium. In due course, it is anticipated that encounters with non-material waves will transpire, and the prior exposure to perturbations in material waves will prove to be an indispensable source of guidance (Mogensen et al., 2014).

The phenomenon, referred to as a disturbance or oscillation, has the potential to propagate uniformly across the medium, exhibiting periodicity at every point, which is known as a periodic wave. Alternatively, the oscillations may be transient, occurring for a finite duration at each point, resulting in a wave pulse. For instance, when a single end of a rope, as illustrated in Figure 1.1, is subjected to shaking, it propagates a solitary wave pulse along the length of the rope. Conversely, a periodic wave can be generated on the rope by persistently and regularly oscillating the rope (Dietz, 1963).

The initiation of the waves in a medium is facilitated by external forces that influence select particles within the medium, thereby inducing oscillatory motion. The entity responsible for producing these external forces is called the wave source. The perpetuation of a wave does not necessitate the presence of external forces once it has been initiated. Let us contemplate a water wave propagating along the surface of a pond. The initiation of wave motion is attributed to an external perturbation: a rock being dropped into the pond. Once initiated, the ripples propagate outward autonomously. Formal Definition of Waves: Formulating a comprehensive and universally applicable definition for "wave" presents a challenging endeavor. Currently, we will suffice by documenting the subsequent elucidation of a material wave (Creze et al., 2018):

A material wave refers to the propagation of a disturbance within a medium, characterized by the collective motion of particles constituting the medium, which undergo oscillations around a stable equilibrium position (Fernández et al., 2013).

Let us make another attempt to comprehend the precise nature of wave motion by elucidating it through the use of an illustrative example (Figure 1.2). Let us contemplate the propagation of ripples emanating from a disturbance within a body of water. Although a certain degree of motion exists among the individual molecules of water, their movement is limited to vertical oscillation and lacks significant horizontal displacement (Faber et al., 2006). The ripples cause significant movement. The phenomenon referred to as a wave is characterized by the propagation of ripples along the surface of a body of water. The phenomenon commonly referred to as a wave is not solely attributed to the motion of individual water molecules or the disruption of the water surface from its equilibrium state. Rather, the propagation of this disturbance to various locations within the medium gives rise to what we define as a wave (Hirsch et al., 2019).

To understand the distinction between types of motion in wave descriptions, one can conduct a simple experiment within the confines of their home. To experiment, obtain a bowl filled with water and carefully introduce a small quantity of olive oil onto the water's surface, creating a configuration akin to the illustration provided. Does the act of gently oscillating one's hand at position x result in ripples at position y? What is the current status of the oil industry? Does it transition to the designated position y (Gregory et al., 2019)?

Figure 1.2. Wave motion experiment (Source: Science Buddies, Creative Commons License).

1.2.1. Pulse Wave

A pulse refers to a perturbation characterized by a finite duration. For instance, shaking the end of a rope once would generate a pulse wave, as depicted in Figure 1.3.

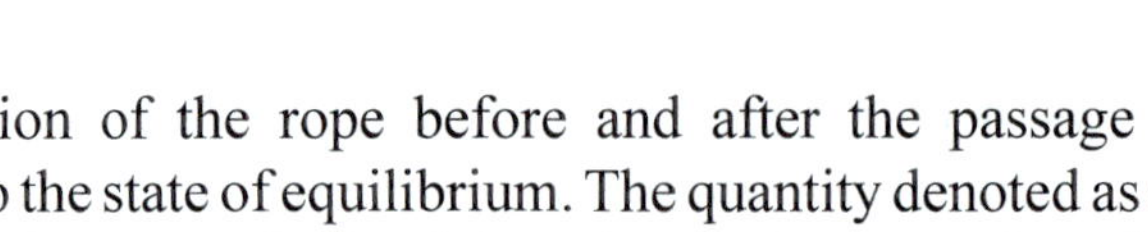

Figure 1.3. Wave pulse (Source: School for Champions, Creative Commons License).

The position of the rope before and after the passage of the pulse corresponds to the state of equilibrium. The quantity denoted as amplitude, A, refers to the highest magnitude of the pulse displacement from its equilibrium position. The variable A is defined to have a positive value regardless of whether the displacement is below or above the equilibrium dashed line level depicted in the figure. The amplitude of a wave is contingent upon the source, which refers to the quantity of energy imparted to the medium by an external entity, such as physically displacing a rope by hand. An additional illustration of a pulse-type wave can be observed in the scenario of a rock being cast into a body of water. In this particular instance, the rock generates multiple ripples; however, they can be classified as pulse waves due to their finite nature, characterized by a distinct initiation and termination (Hirata et al., 2006).

1.2.2. Periodic Wave

Recurrent undulations can exhibit a diverse array of forms. The sine or cosine function is one of the most straightforward functions to manipulate. These types of waves are commonly referred to as harmonic or sinusoidal waves. The generation of these oscillations is facilitated by oscillators that exhibit simple harmonic motion akin to the spring-mass system examined in the context of 7A. To clarify, the behavior of a harmonic wave can be represented by the motion of individual particles within the medium, which can be described using a spring-mass oscillator model. When an individual grasps one end of a rope and oscillates it in a regular up-and-down motion, known as simple harmonic motion, a series of harmonic waves is produced. If a visual representation of the wave were captured as a photograph, it would resemble the Figure 1.4 (Nakamura, 1979).

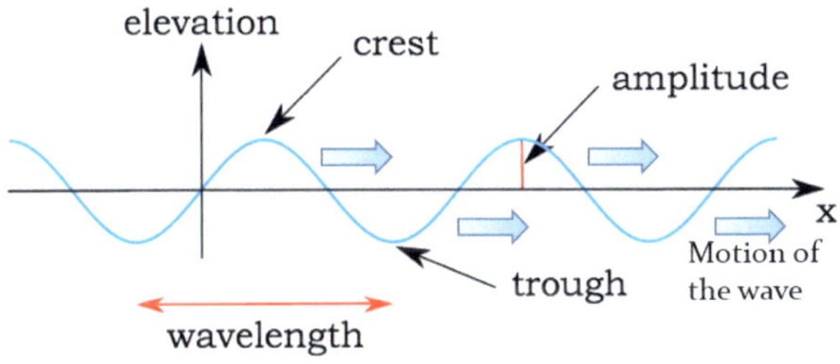

Figure 1.4. Representation of a periodic wave (Source: IOWA State University, Creative Commons License).

When an image is captured of a wave, it represents its form at a specific moment in time, commonly referred to as the snapshot of the wave. The graphical representation of this can be observed as depicted in Figure 1.5. In this context, the variable x denotes the spatial position along the rope, while the variable y represents the displacement of the rope from its equilibrium position (Cordesses, 2004).

Like the wave pulse, a periodic wave possesses a clearly defined state of equilibrium and an amplitude denoted as A. The equilibrium position in Figure 1.4 is at y=0 cm, while the amplitude measures 2 cm. In contrast to a wave pulse, a repeating wave exhibits the presence of two additional parameters. One of the fundamental properties of a wave is its wavelength, denoted as λ. This characteristic represents the minimum distance, measured in the direction of wave propagation, between successive identical points of the wave. The wavelength denotes the spatial extent of the wave's cycle, as indicated in Figure 1.4, presented earlier. In the depicted diagram, the symbol λ represents a value of 4 meters (Wang & Li, 2005).

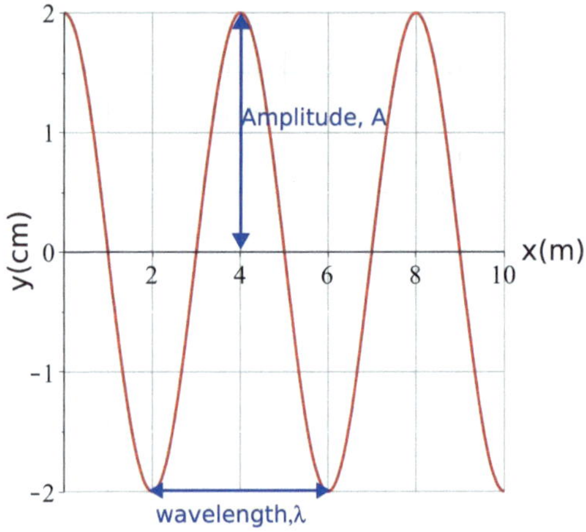

Figure 1.5. Graphical representation of a harmonic wave at a fixed time (Source: The California State University, Creative Commons License).

The second novel parameter pertains to the period, denoted as T, which represents the duration required for the wave to exhibit identical characteristics. But, the wavelength provides information about the spatial repetition of a wave, whereas the period provides information about the temporal repetition of a wave. The duration of one period is equivalent to the time required for a wave to propagate a distance equal to one wavelength, as it will exhibit identical characteristics upon completion of a single cycle (Nakamura, 1980). Oscillatory systems within a medium complete a single cycle within a single period, exemplified by reversing a spring-mass system to its initial position. Pulse-type waves, characterized by their non-repetitive nature, lack distinct periods or wavelengths. The frequency at which the medium is disturbed by the source determines the period of the wave. When generating a harmonic wave on a rope, the period can be determined by measuring the duration for one complete cycle of upward and downward movement of the hand (Pu et al., 2020). The graph depicted in Figure 1.4 illustrates a wave at a specific moment in time, thus providing no insights into the time-dependent nature of its period. To depict the temporal attributes of the wave, it is necessary to create a revised graphical representation. To depict the temporal evolution of a wave, it is necessary to fix its position by representing it in terms of space and time, as these two variables collectively describe the wave. If a red dot is painted on the rope at a specific x value and its position is plotted against time, it would reveal the motion of that particular point in simple harmonic motion (refer to Figure 1.6). Hence, when expressing the temporal evolution of the wave, we are illustrating the harmonic movement of a particular particle within the medium, situated at a constant spatial position. The temporal duration for the "dot" to complete a full cycle and return to its original position is equivalent to the time required for the entire wave to complete a full cycle and return to its initial configuration, as visually depicted in Figure 1.6. A representation of a displacement as opposed to a time graph is depicted in Figure 1.6 (Zhou et al., 2003).

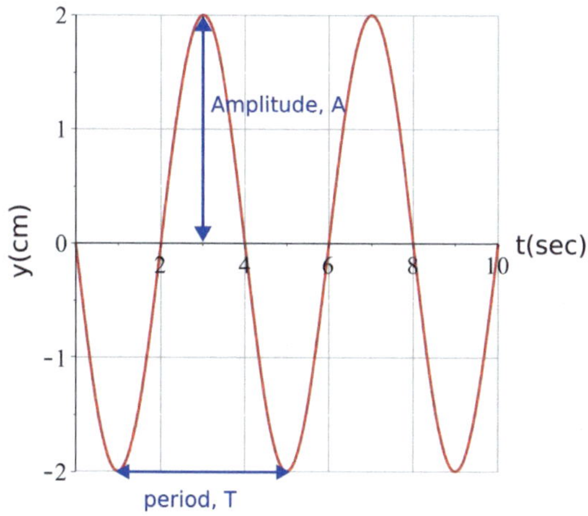

Figure 1.6. Graphical representation of a harmonic wave at a fixed position (Source: The California State University, Creative Commons License).

The graph above depicts a temporal cycle with a duration of 4 seconds. Given the assumption that Figures 1.5 and 1.6 depict identical waveforms, the data conveyed by these figures must exhibit congruence. Both graphs exhibit an identical magnitude of 2 cm. The wave exhibits a maximum displacement, called the crest, with a magnitude of 2 cm. Conversely, the wave demonstrates a minimum displacement, known as the trough, with a magnitude of -2 cm (Liu et al., 2001). The equilibrium position at y=0 cm represents the middle point between the crest and the trough. Figure 1.5 illustrates the positioning of the particle at various locations along the x-axis. Specifically, it shows that when the particle is situated at x=0 m, it is at a crest. At x=1 m, the particle is positioned at equilibrium; at x=2 m, it is situated at a trough. Let us assume that Figure 1.5 represents a snapshot taken at t=0 seconds. Given that Figure 1.6 depicts the state of equilibrium at t=0 sec, the corresponding position in Figure 1.5 must also be at equilibrium. Therefore, potential locations for the time-varying graph in Figure 1.6 may include x=1, 3, 5, and so on, where the displacement reaches equilibrium, as depicted in Figure 1.5 (Shen et al., 2021).

The period refers to the duration required to complete a single cycle. The frequency, which refers to the number of cycles that can be accommodated within a given time, is an additional valuable indicator of wave periodicity. The inverse of the period is referred to as frequency (Ma et al., 2009).

$$T = \frac{1}{f} \tag{1.1}$$

In general, the duration of a period is quantified in units of seconds, while the frequency is quantified in Hertz (Hz), where 1 Hz is equivalent to 1 s-1. The period of a wave refers to the temporal duration between the successive arrival of neighboring crests. In contrast, the frequency denotes the number of crests trailing a given point within a unit of time, typically measured in seconds. This implies that the red dot depicted in Figure 1.6 undergoes a specific number of complete up-and-down cycles within a second. Alternatively, when considering a particular spatial location, a wave will cyclically revert to its initial configuration several times within a second (Yan & Zhang, 2001).

1.2.3. Dimensionality

The instances mentioned above are called one-dimensional waves due to their characteristic of propagating solely in a single direction. The designation of that particular direction as the x-axis was made arbitrarily. If a wave propagates across a surface, it can be classified as a two-dimensional wave. An illustration of this concept can be observed in the form of ripples that manifest on the surface of a pond, which can be understood as a manifestation of a two-dimensional water wave. A wave that propagates in a spherical manner, expanding equally in all directions, can be classified as a three-dimensional wave. Typical examples of three-dimensional waves include sound waves and light waves (Grilli et al., 2001). One notable differentiation among these waves is that the amplitude A of the waves remains constant solely for one-dimensional waves. The amplitudes of two-dimensional and three-dimensional waves are contingent upon the spatial proximity to the origin of the perturbation. The phenomenon of ripples in a pond exhibits a reduction in size as they propagate outward. At the same time, the amplitude of light waves, representing brightness, and sound waves, representing loudness, diminishes as the distance from their respective sources increases. The observed phenomenon can be attributed to the principle of energy conservation. As a wave expands in multiple dimensions, its energy is inevitably distributed across a progressively larger area (Cally & Goossens, 2008).

1.2.4. Polarization

Polarization is a distinctive attribute exhibited by both material waves and electromagnetic waves. The concept of polarization establishes a relationship between the direction of wave displacement and the direction of wave motion. In the context of material waves, it is necessary to establish two distinct forms of polarization (Dwivedi et al., 2021).

1.2.4.1. Transverse Polarization

A transverse material wave is characterized by a displacement from equilibrium that is perpendicular to the direction of wave propagation. The observed wave can be classified as a transverse wave due to its rightward propagation, while the oscillations within the medium occur in a vertical direction perpendicular to the wave's motion. The oscillatory motion of a single oscillator can be symbolized by the "red dot," which exhibits periodic upward and downward movement. This motion occurs as the wave propagates through the medium in a rightward direction. One illustration of a transverse wave can be observed as a wave produced on a rope or string, as depicted in Figure 1.7. A surface water wave can be considered an additional illustration, as the vertical oscillations of the water particles occur while the wave propagates within the two-dimensional plane of the water surface (Chang et al., 1986).

Figure 1.7. Transverse wave (Source: Alex Ayres, Creative Commons License).

1.2.4.2. Longitudinal Polarization

A longitudinal material wave is characterized by the displacement of the medium from its equilibrium position being aligned in a similar direction as the wave's propagation (Figure 1.8). In most instances involving longitudinal waves examined, the displacement manifests as periodic compressions, which correspond to regions of higher density within the medium, and rarefactions, which correspond to regions of lower density within the medium. The diagram below depicts an illustrative instance of a longitudinal wave propagating along a spring (or slinky). Longitudinal waves, such as sound waves, represent the most prevalent instance that will be further elaborated upon in a subsequent section (Zheng et al., 2019).

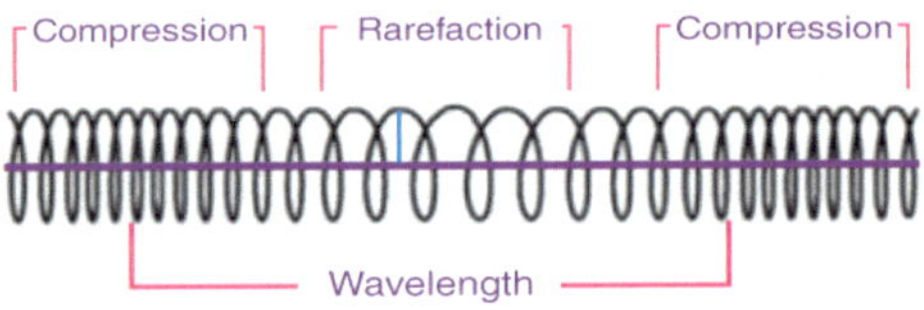

Figure 1.8. Longitudinal wave (Source: BYJU's, Creative Commons License).

While transverse waves resemble the plots depicted in Figures 1.5 and 1.6, harmonic longitudinal waves are represented similarly using sinusoidal functions. The representation of displacement on the y-axis in Figures 1.5 and 1.6 does not necessarily correspond to the physical depiction of the wave. Instead, it indicates the distance from equilibrium within the medium. Similar to transverse waves, the red dot in longitudinal waves exhibits sinusoidal oscillation. The locations where compressions occur can be understood as crests, while the locations where rarefactions occur can be understood as troughs (Habib et al., 2012).

1.2.5. Wave Speed

The velocity of a wave, denoted as a wave, represents the rate at which the perturbation travels within the given medium. One perspective regarding wave speed is that it represents the velocity at which an individual riding a surfboard on the wave would move. The velocity of the oscillations exhibited by the constituent particles comprising the medium is not the determining factor. According to the findings in 7A, it is evident that the velocity of a harmonic oscillator, specifically a spring-mass system, is not a consistent value but rather varies to the displacement from the equilibrium position. The velocity of the oscillator is at its maximum when it is in a state of equilibrium and subsequently diminishes as the oscillator deviates further from this state, ultimately reaching zero when it reaches its maximum displacement from equilibrium (Khir et al., 2001).

Tip

Understanding wave speed is crucial as it determines how fast a wave propagates through a medium. It is influenced by the properties of the medium, such as its density and elasticity.

To a reasonable approximation, the velocity of a wave is primarily influenced by the characteristics of the medium through which it propagates rather than being dependent on the amplitude or frequency of the wave. The validity of this approximation is compromised when dealing with large waves or waves characterized by extreme frequencies. Currently, we will simplify our discourse by disregarding the influence of amplitude on wave speed. This is due to our focus on smaller waves in the context of 7C. The consideration of the dependence of speed on frequency will be deferred until the refraction of light waves is addressed in a subsequent chapter (Hudson, 1981).

The definition of speed, as stated in the seventh-grade science curriculum (7B), can be expressed as follows:

$$\text{Speed} = \frac{\textit{distance travelled}}{\textit{Time Spent}} \tag{1.2}$$

The minimum duration required for a wave to exhibit complete replication of its initial appearance is equivalent to one period. By definition, when the wave exhibits identical characteristics once more, it has traversed a distance equivalent to one wavelength. The wave speed can be expressed as the ratio of the distance traveled by a wave, which is equal to one wavelength, to the time it takes for the wave to complete one full period (Feng & Khir, 2010).

$$v_{wave} = \frac{l}{T} = \lambda f \tag{1.3}$$

The second equality mentioned above employs the meaning of frequency as stated in Eq. (1.1). An illustration of how the medium influences the speed of a wave can be observed in the case of a material wave propagating through a stretched medium, such as a rope or a rubber hose, which is characterized by tension (Dawson & Elliott, 1977). Transverse waves and longitudinal waves can both occur on taut strings or ropes. The velocity denoted as v_{wave}, of transverse waves propagating along a stretched string, is contingent upon the characteristics of the string that influence its elasticity and internal properties. The relationship between the wave speed and the properties of a string, when the string is thin in comparison to its length, can be expressed as follows (Vinh & Ogden, 2004):

$$v_{wave} = \sqrt{\frac{T}{\mu}} \quad (1.4)$$

In this context, T represents the tension experienced by the string, while μ denotes its mass density, defined as the mass per unit length ($\mu = m/L$). Tension refers to the force exerted by one string upon another. The relationship between string tightness and tension is such that an increase in string tightness leads to a corresponding increase in tension. As previously discussed, a material wave refers to a perturbation propagating through a medium utilizing intermolecular forces between adjacent particles. It follows that an augmentation in tension within the medium leads to a corresponding augmentation in the velocity of the wave. An increase in the magnitude of forces will result in a corresponding increase in the rate of acceleration. When the mass of the string is relatively high, the intermolecular forces between different segments of the string lead to reduced acceleration (Rosakis et al., 1999).

Consequently, it is also reasonable to expect that as the coefficient of friction (μ) increases, the speed of the wave will decrease. The manipulation of wave speed is important in stringed instruments such as the guitar or violin. This is evident through tuning knobs at one end, which allow tension adjustment and using strings with varying mass (represented by μ) to achieve different wave speed values. A more comprehensive examination of string instruments will be undertaken in subsequent sections, specifically when addressing the concept of standing waves (Ablowitz & Zeppetella, 1979).

1.3. WAVE REPRESENTATION

The concept of wave representation pertains to the various methods by which waves can be mathematically or visually illustrated and characterized. Graphs are frequently employed to visually represent waves, wherein the vertical axis denotes the magnitude of the wave (referred to as the amplitude). In contrast, the horizontal axis denotes the passage of time or the extent of distance. The graphical representation facilitates the observation of the wave's morphology, oscillatory behavior, and pertinent attributes such as amplitude, wavelength, and frequency. Mathematical equations describe waves, facilitating a more accurate comprehension of their temporal dynamics. An example of a sinusoidal wave can be mathematically expressed as $y = A * \sin(2\pi ft + \varphi)$, where A denotes the amplitude, f represents the frequency, t signifies time, and φ denotes the phase angle (Besieris et al., 1989).

Waveform displays are commonly found in electronics and signal processing, where they serve as visual representations of signals that change over time. These signals can include audio signals as well as electrical signals. The displays mentioned above depict the magnitude of the wave as a function of time, thereby offering a visual depiction of the fluctuations in the signal. Phasor diagrams, conversely, serve as visual representations that depict the phase correlation among multiple waves. Phasor diagrams facilitate the comprehension of wave interference and the resultant constructive or destructive interference by employing vectors (phasors) to depict each wave's amplitude and phase angle (Devaney & Wolf, 1974).

Wavefronts are theoretical surfaces that connect all points of a wave in phase, illustrating the spatial propagation of waves. In three-dimensional representations, wavefronts manifest as surfaces, whereas in two-dimensional representations, they are depicted as lines. Spectrograms are visual depictions that illustrate the frequency characteristics of a waveform to time. Frequency spectrums are commonly employed in audio and signal processing to visually represent the various frequency components in sound or other signals that change over time (Suppes & Han, 2000).

Vector diagrams are commonly employed to represent the orientation of the magnetic and electric fields for the direction of propagation for transverse electromagnetic waves. The representation of waves is of utmost importance in comprehending waves' behavior, propagation, and interaction with different mediums. Using these visual and mathematical methodologies, scientists and engineers can examine waves, develop wave-centric technologies, and formulate forecasts regarding wave phenomena across various domains, encompassing telecommunications, acoustics, optics, and more (Gouesbet et al., 1995).

1.3.1. Simple Harmonic Motion

The subsequent portion of the course will be dedicated to examining harmonic waves, a distinct type of infinite repeating waves. These waves were previously elucidated in the preceding section and will now be the primary focus of our study. In the preceding discussion, we examined the graphical representation of waves and established their fundamental associated characteristics. In this section, we aim to formulate a mathematical expression that describes one-dimensional harmonic waves and establish the relationship between the mathematical representation and the graphical depiction of these waves (Galeriu et al., 2014).

Remember

Simple Harmonic Motion (SHM) is a fundamental concept in wave representation. Many waves, such as sinusoidal waves, can be described using SHM, where the wave oscillates back and forth around an equilibrium position.

To mathematically represent the wave, it is necessary to consider a single oscillator within the medium. In the preceding section, it was established that when a wave propagates through a medium, the movement of individual particles within the medium can be regarded as that of an oscillator centered around the equilibrium position of the medium. When the wave exhibits harmonic behavior, its oscillations can be characterized by a simple harmonic oscillator akin to the spring-mass system examined in the 7A course, as illustrated in the Figure 1.9 (Somroob & Wattanakasiwich, 2017).

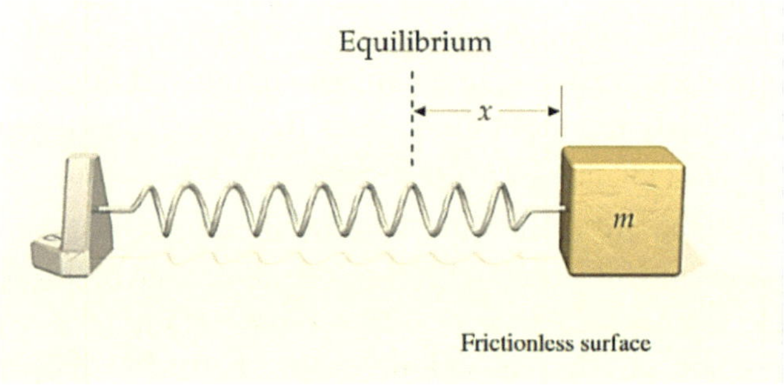

Figure 1.9. Simple harmonic oscillator (Source: Go Seeko, Creative Commons License)

Our primary objective was to comprehensively describe the energy associated with the harmonic oscillator, specifically in terms of its potential and kinetic components compared to the equilibrium position. The present objective is to direct our attention toward the dynamics of the harmonic oscillator and to elucidate the temporal evolution of the oscillator's position. The above outcome can be derived using Newton's second law in conjunction with the force equation for a simple harmonic oscillator. Hooke's Law elucidates the restorative force governing the behavior of simple harmonic motion (Triana & Fajardo, 2011).

$$F_{net} = -ky \quad (1.5)$$

In the given context, the variable "k" represents the spring constant, which characterizes the stiffness of the spring. On the other hand, "y" denotes the displacement from the equilibrium position. A comprehensive understanding can be achieved by integrating Hooke's Law with Newton's second law (Kinchin, 2016).

$$F_{net} = ma = -ky \quad (1.6)$$

By utilizing the concept of acceleration and rephrasing the equation mentioned above in a manner that allows for the determination of position as a function of time, the resulting expression can be derived as follows:

$$md^2y/\ dt^2 + ky = 0 \quad (1.7)$$

Based on the equation mentioned above, it is evident that the variable y must be expressed as a function that, upon being differentiated twice, yields the negative of the original function. The functions that possess this characteristic are sinusoidal, specifically sine or cosine. The representation of the position of a harmonic oscillation as a function of time can be derived as a consequence (Afandi, 2018).

$$y(t) = A\sin(\omega t+\phi_o) \quad (1.8)$$

where $\omega = \sqrt{k/m}$. Representing periodic harmonic motion using a sine function that exhibits repetition is justifiable. The amplitude, denoted as A, represents the maximum displacement of a sine function, as the function's values are bounded by a maximum value of one. In the context of the spring-mass system, the amplitude denotes the magnitude of the displacement from the equilibrium position that was imparted to the system before its release. In accordance with the principle of energy conservation, our findings in the 7A study indicate that the displacement of the spring-mass system cannot exceed the initial distance from which it was released (Pambayun et al., 2019).

The constant ϕo, when placed within the sine function, determines the system's initial displacement at t=0 seconds. For instance, in the scenario where the spring-mass system is initially displaced downwards by a distance of y = -A at time t = 0 seconds, the following observations can be made:

$$y(0) = -A = A\sin(\phi_o) \Rightarrow \sin(\phi_o) = -1 \Rightarrow \phi_o = \frac{3\pi}{2} \quad (1.9)$$

The sine function, as described by Eq. (1.8), is graphed below for the given value of $\phi o = 3\pi/2$. The graph begins with the highest displacement below the equilibrium position (Lan & Shih, 2012).

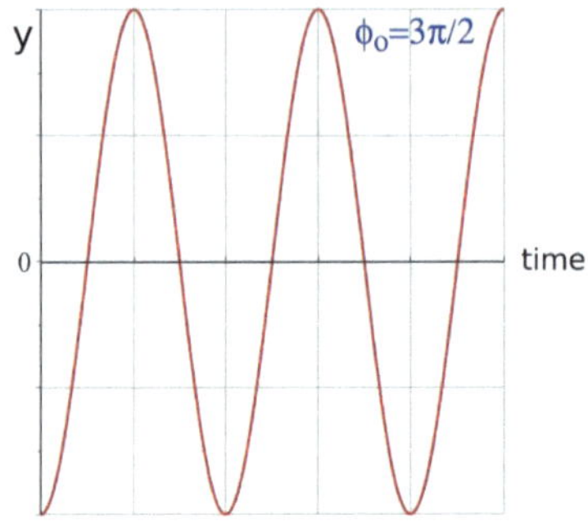

Figure 1.10. Phase constant example (Source: The California State University, Creative Commons License).

The sine function in Eq. (1.8), with the given value of $\phi o = 3\pi/2$, is depicted in the graph above, commencing from a maximum amplitude. In an alternative scenario, when the initial conditions (t=0s) are defined when the spring-mass system crosses the equilibrium position (y=0), the maximum displacement below the equilibrium can be determined.

$$y(0) = 0 = A\sin(\phi_o) \quad (1.10)$$

The scenario mentioned above presents a higher level of complexity due to two solutions within a single cycle of the sine function for the given initial scenario, namely $\phi o = 0$ or $\phi o = \pi$. The phase constants are associated with two distinct physical scenarios occurring during a single cycle when the mass is in equilibrium. If the mass is oriented vertically, as depicted in Figure 1.11, it can move in either a downward or upward direction as it traverses the equilibrium point. The provided figure illustrates that when the spring-mass system is in equilibrium and moving upwards, it corresponds to a phase angle (ϕo) of 0. Conversely, when the system is in equilibrium and moving downwards at t=0 seconds, the phase angle is π. The left figure is shifted by half a cycle to obtain the right figure (Dimas et al., 2018).

According to the definition of a period T, the function is expected to exhibit a return to its initial position after completing one full cycle. In trigonometric functions, radians are a practical unit for measuring the number of cycles.

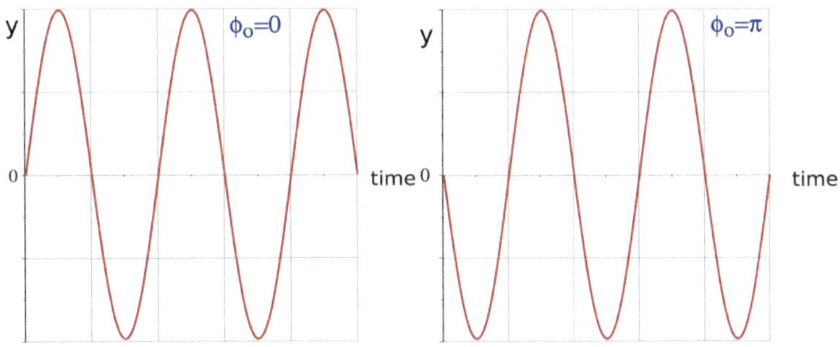

Figure 1.11. Phase constant (Source: Dina Zhabinskaya, Creative Commons License)

The angular measure of 2π radians corresponds to a complete revolution or one cycle, such that:

$$\sin(x + 2\pi) = \sin(x) \quad (1.11)$$

To ensure that the argument of the sine function is expressed in radians, Eq. (1.8) can be reformulated as follows: (1.11)

$$x(t) = A\sin(\ 2\pi/T\ t + \phi o) \quad (1.12)$$

The equation ω Eq. (1.8) can be expressed as $2\pi/T$. To verify the consistency of this expression with the definition of one period, it is necessary to ensure that the function remains unchanged when a time, denoted as $t \rightarrow t+T$, elapses (Amrani & Paradis, 2010).

$$y(t+T) = A\sin\left(\frac{2p}{T}(t+T)+f_o\right) = A\sin\left(\frac{2p}{T}t+2p+f_o\right) = A\sin\left(\frac{2p}{T}t+f_o\right) = y(t) \quad (1.13)$$

1.3.2. Harmonic Wave Equation

The preceding discourse, which culminated in the derivation of Eq. (1.12), fails to encompass the entirety of wave phenomena, as it solely characterizes the temporal evolution of a solitary oscillator situated at a stationary location within the medium. The term "fixed position" pertains to the specific point within the medium around which the particle oscillates. It is necessary to incorporate the spatial wave motion within the medium to provide a comprehensive description of the wave. But, our objective is to formulate a mathematical expression that accurately represents the displacement of every oscillator within the medium at any given moment. The mathematical representation of a one-dimensional harmonic wave can be expressed as follows (Riyanti et al., 2006):

$$y(x,t) - y_o = A\sin\left(\frac{2p}{T}t \pm \frac{2p}{l}x + f_o\right) \tag{1.14}$$

The expression (x,t)−yo represents the displacement from equilibrium at a given position x and time t. In this equation, yo represents the equilibrium position, while y(x,t) denotes the oscillator's position relative to a selected reference point. In most instances, it is advantageous to establish the initial position, yo, as zero. This choice yields a symmetrical displacement pattern around the x-axis, with the maximum displacement occurring at point A and the minimum displacement at point -A. However, in cases where the equilibrium position is not defined as zero, the sine function undergoes a vertical shift, either upward or downward Colson, 1981).

Equation (1.12) illustrates that in the case of a solitary harmonic oscillator, the sine function retains its identical form following the passage of one complete period. The second term in Eq. (1.14), which characterizes the spatial periodicity of the wave, exhibits a strong resemblance to the first term. By employing the identical approach as demonstrated in Eq. (1.13), it can be demonstrated that the wave equation remains unchanged when any integer multiple of the wavelength increments the value of x (Fliss & Joly, 2016).

The fixed phase constant ϕo differs from the one stated in Eq. (1.12) as it characterizes the waveform at both the initial time, t = 0, and the reference point x = 0. As an illustration, let us consider a scenario where a crest is present at the coordinates x = 0 and t = 0. Assuming the equilibrium position is zero (yo = 0), the initial displacement of the wave can be identified as A, as it corresponds to a crest (Nédélec, 2001).

$$y(0,0) = A = A\sin(\phi_o) \Rightarrow \sin(\phi_o) = 1 \Rightarrow \phi_o = \frac{p}{2} \tag{1.15}$$

The symbol ± prompts the selection of either the positive (+) or negative (−) sign, contingent upon the direction in which the wave propagates. Determining the appropriate sign to the direction of wave propagation, specifically when the wave propagates is moving towards the right (with increasing values of x) or the left (with decreasing values of x), is of interest. The variable y(x,t) represents the displacement of a specific wave component as it propagates in both spatial and temporal dimensions. If one observes a specific segment of a wave, the displacement at that segment should remain constant regardless of variations in position and time. Consider a scenario where one begins observing a crest at time t = 0 of a wave that originates from the position x = 0 and propagates towards the right. Suppose one observes the crest for a duration equivalent to one-fourth of a period, denoted as t = T/4. In that case, this crest will be located at a position corresponding to one-fourth of a cycle in the positive x direction, specifically x = λ/4. By utilizing the outcome presented in Eq. (1.15), which pertains to the phase constant in situations where the initial circumstances of the wave correspond to a crest, one can determine the displacement at a specific time t = T/4 (Guo & Guo, 2013).

$$y\left(\frac{l}{4},\frac{T}{4}\right) = A\sin\left(\frac{2p}{T}\cdot\frac{T}{4} \pm \frac{2p}{l}\cdot\frac{l}{4} + \frac{p}{2}\right) = A\sin\left(\frac{p}{2} \pm \frac{p}{2} \pm \frac{p}{2}\right) = \left\{\begin{array}{l} + \Rightarrow A\sin\left(\frac{3p}{2}\right) = -A \\ - \Rightarrow A\sin\left(\frac{p}{2}\right) = A \end{array}\right\} \quad (1.16)$$

Given that the crest was being tracked, it is appropriate to select the negative sign, as it led to a displacement equivalent to A, as indicated by the equation mentioned above. When the wave propagates towards the left, its position after a quarter period would be at x = -λ/4. Following the procedure mentioned above, it can be deduced that a positive sign preceding the spatial term would yield an accurate outcome. In summary, the convention is to assign a "+" sign to denote left-moving waves and a "–" sign to denote right-moving waves when expressing the wave equation as presented in Eq. (1.14) (Kuvshinov & Mulder, 2006).

There exists an additional highly advantageous measure that we will frequently employ to characterize wave characteristics. The total phase, denoted as Φ, should not be mistaken for the fixed phase constant ϕo. It is precisely the value enclosed within the sine function of the wave equation (Ali et al., 2020).

$$\Phi(x,t) \equiv 2\ddot{U}\,\frac{t}{T} \pm 2\;\;\frac{x}{l} + \;\;{}_{o} \quad (1.17)$$

The concept of the total phase refers to the specific location within a wave. When individuals engage in the mental exercise of envisioning themselves surfing a wave or observing the movement of a wave's peak or trough, they are essentially tracking a fixed point of uninterrupted phase. Returning to the topic of directional discourse, it is imperative to maintain a constant phase for any segment of a wave (Brambilla, 2002). Consequently, if the wave is propagating towards the right, with both time and position exhibiting an increase, it is necessary for the sign preceding the spatial term to be negative to uphold the constancy of Φ. In contrast, the sign must be positive when the wave exhibits leftward movement (with time increasing while position decreases) to maintain a constant total phase (Garfinkle, 2002).

Given the complexity of the wave equation, it is pertinent to acknowledge the distinction between variables and parameters, as well as the interdependence of parameters on the wave source and the medium of propagation.

- The constants A, T, λ, ϕo, yo, and the choice of + or – remain constant for a given harmonic wave. The authors describe the wave and its corresponding behavior. The determination of a wave's period and amplitude is contingent upon its source's characteristics. Determining the phase constant, equilibrium position, and direction is contingent upon selecting an origin and coordinate system, yet once established, these parameters remain constant. The velocity of a wave is contingent upon the properties of the medium through which it propagates. The wavelength is contingent upon the wave's period and velocity.
- The variables, denoted as x and t, encompass the entirety of the spatial and temporal dimensions in which the wave is present. These variables determine the displacement of the medium at a specific time and location.
- The investigation of displacement at various locations and times can be conducted by manipulating variables, whereas parameters associated with a wave remain constant values (Kress, 1979).

Having derived an equation that comprehensively characterizes one-dimensional wave phenomena, we aim to ascertain the feasibility of visually representing this equation through graphical means. Visualizing a particular behavior involves representing the equation describing it as a mathematical function concerning the relevant variable, and plotting this function. However, in the context of the wave equation, two variables exist space and time. One can establish a specific time and generate a graphical representation of the wave about its position. This graphical representation, the snapshot plot in the preceding section, solely depicts the wave's spatial characteristics and lacks any temporal information. Alternatively, one may choose to observe a specific location within the medium and analyze the temporal oscillations of that particular segment. However, this approach entails sacrificing the ability to capture the spatial characteristics of the wave. Hence, it becomes evident that a comprehensive depiction of wave phenomena necessitates the inclusion of multiple plots. However, can only two plots adequately capture the wave's description at all positions and instances of time? Let us commence our analysis by examining the two plots depicted herein (Adelman & Doll, 1976).

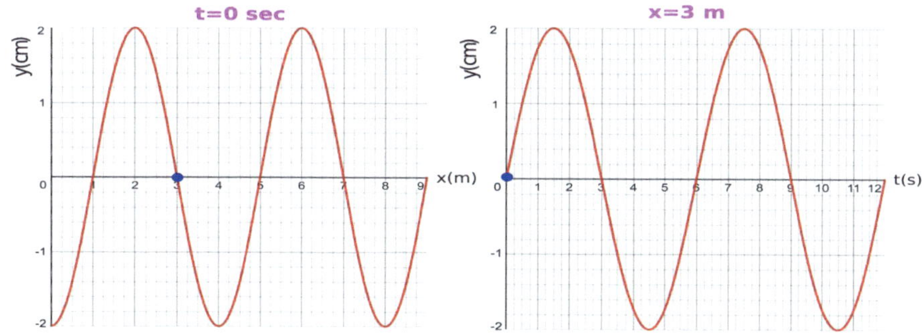

Figure 1.12. Graphical representation of a wave (Source: Dina Zhabinskaya, Creative Commons License).

The plot on the left represents a snapshot of the wave at t = 0 seconds. The plot provides information regarding the wave amplitude, denoted as A, which has a value of 2 cm, and the wavelength, represented by λ, which measures 4 m. The plot on the right illustrates the oscillatory behavior of the medium at a specific spatial location, specifically x = 3 m, as a function of time. The plots must exhibit consistent amplitudes. The plot additionally presents the time interval data, T = 6 seconds. Both plots demonstrate that the equilibrium point is established at zero, specifically yo = 0 (Deng, 2003).

To comprehensively articulate the wave equation, it is imperative to ascertain both the direction of the wave and the constant phase that remains fixed. Let us first consider the concept of direction. A snapshot of the moving wave is provided to us at time t = 0 seconds. We aim to ascertain whether the available data from the two plots is sufficient to determine the subsequent direction of wave shift, either towards the right or the left, upon resuming the "movie." The sole supplementary data indicates that the displacement at the position x = 3 m will increase after t = 0 sec, implying the propagation of a crest towards the x = 3 m location. The two dots in Figure 1.12 correspond to identical spatial and temporal coordinates (t = 0 s and x = 3 m). To clarify, the two dots represent the shared point of intersection between the two plots, denoting an equilibrium displacement.

Given that the right plot indicates a positive displacement at x = 3 m after t = 0 s, aligning the snapshot plot in the appropriate direction of motion should yield an equivalent outcome. Given our objective of positioning a crest at the x = 3 m location, shifting the snapshot plot towards the right is necessary, which illustrates the shifted snapshot at a time interval of 0.5 seconds.

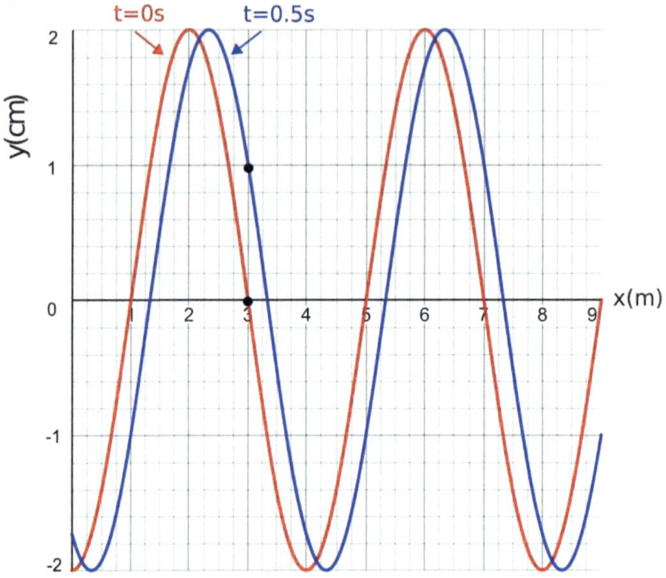

Figure 1.13. Shifting the snapshot plot (Source: UC Davis, Creative Commons License).

According to the time plot, it can be observed that at t = 0.5 s, the displacement at x = 3m is estimated to be approximately 1 cm. The duration of 0.5 seconds corresponds to one-twelfth of a complete cycle, given that a complete cycle lasts 6 seconds. This phenomenon can be observed in the Figure 1.13, where the two overlapping plots demonstrate a displacement corresponding to approximately 1/12 of a wavelength or 1/3 m. Following the transition, the displacement at the position x=3m increased by 1 cm, aligning with the anticipated outcome. If the plot were to be shifted to the left, the displacement would exhibit a decrease, which is incongruent with the temporal nature of the plot. Hence, it can be inferred that the observed wave propagates in the right direction (Shao & Faltinsen, 2014).

Finally, let us examine the feasibility of determining the fixed phase constant based on the information presented in these two plots. The variable ϕo is designated as the initial displacement of the wave at the specific time t = 0 s and position x = 0 m. Substituting the given values into Eq. (1.14) yields:

$$y(0,0) = -2\text{cm} = (2\text{cm})\sin(\phi_o) \Rightarrow \sin(\phi_o) = -1 \Rightarrow \phi_o = \frac{3p}{2} \qquad (1.18)$$

The acquisition of the phase constant was straightforward. Nevertheless, extracting all values of the fixed phase constant from the plot is not always straightforward. As an illustration, an established phase constant of $\pi/4$ corresponds to a displacement of 0.7071 (assuming an amplitude of one). This particular value may pose challenges in terms of interpretation when examining the plot. However, it is unnecessary to rely solely on the values at the origin and initial time. One can select any point on the plot and determine the phase constant by solving. The graphical representation provided in Figure 1.12 shows that at a position of x = 2 m, the displacement corresponds to a crest. Given that the snapshot plot is taken at time t = 0 seconds, we can substitute this value into the wave equation, utilizing the previously established parameters, resulting in the following outcome (Luostari & Huttunen, 2009):

$$y(x=2m,t=0s) = 2cm = (2cm)\sin\left(\frac{2p}{6s}(0) - \frac{2p}{4m}(2m) + f_o\right) = (2cm)\sin(-p+f_o) \Rightarrow \sin(-p+f_o) = 1 \qquad (1.19)$$

For a crest to occur, the total phase must be equivalent to $\pi/2$, as the sine of $\pi/2$ equals 1. Hence, by solving for the phase constant, we obtain:

$$-\pi + \phi_o = \frac{p}{2} \Rightarrow \phi_o = \frac{3p}{2} \qquad (1.20)$$

This yields an outcome equivalent to the value obtained in Eq. (1.18). The outcome remains consistent for the fixed phase constant, regardless of the alternative location selected from the two plots. In general, selecting either a crest or a trough as a reference point for determining the phase angle ϕo is advantageous. This is because there is a unique value for the total phase at these points, specifically $\pi/2$ for crests and $3\pi/2$ for troughs. An alternative approach is to employ an equilibrium value. However, it is important to note that two potential solutions exist for the total phase, namely 0 or π. Consequently, it becomes necessary to determine the appropriate outcome by considering the function's behavior after the selected equilibrium position. Identifying any value on the plot, aside from the equilibrium, crest, or trough, may pose challenges regarding readability.

An infinite number of solutions exist for the fixed phase constant, as one can manipulate the phase constant by adding or subtracting any multiple of

2π, resulting in an equivalent outcome for the wave equation. For instance, alternative solutions for the phase constant in Eq. (1.18) include -π/2, 7π/2, and 23π/2. Indeed, selecting an alternative crest or trough could yield a value distinct from 3π/2. To maintain simplicity, it is generally advisable to maintain a constant phase value within the range of -π to π.

The wave equation illustrated in Figure 1.12 can be derived by combining all relevant components (Baruch et al., 2009).

$$y(x,t) = (2cm)\sin\left(\frac{2p}{6s}t - \frac{2p}{4m}x + \frac{3p}{2}\right) \tag{1.21}$$

1.4. MULTI-DIMENSIONAL WAVES

Multi-dimensional waves are characterized by their ability to propagate in multiple directions simultaneously. One-dimensional waves propagate along a singular axis, whereas multi-dimensional waves exhibit more intricate spatial patterns. An illustrative instance pertains to waves that exist in two dimensions, such as water waves observed on the surface of a pond. These waves propagate in two perpendicular directions, creating a planar or flat surface. On the other hand, waves in three dimensions, such as sound waves originating from a single point, propagate outward in all directions from the source. This propagation results in pressure fluctuations within the surrounding medium and the formation of spherical wavefronts (Khalid et al., 2021).

Electromagnetic waves exemplify multi-dimensional wave phenomena. Transverse waves, which propagate through three-dimensional space, are observed. The electromagnetic spectrum comprises diverse waves, encompassing radio waves, microwaves, infrared radiation, visible light, ultraviolet radiation, X-rays, and gamma rays. Each wave has specific wavelengths and frequencies, which dictate its distinctive characteristics and practical uses (Raymond & Kuo, 1984).

Seismic waves produced by seismic events such as earthquakes and other geological phenomena exhibit multi-dimensional characteristics. Surface waves and body waves are two types of seismic waves that propagate through the Earth's interior, yielding important insights into the structure and composition of the Earth (Lin et al., 2016).

Acoustic waves in solids can propagate in various dimensions, contingent upon the material's crystal structure and the wave's specific characteristics.

In the context of a crystal lattice, it is observed that sound waves can propagate along multiple directions, thereby giving rise to the phenomenon known as acoustic dispersion (Xu et al., 2021).

1.4.1. Sound Waves

Sound can be understood as a wave, as per our established definition. Nevertheless, the behavior of sound waves in air differs from that of material waves. As previously examined in Physics 7A, it has been established that the intermolecular bonds within the air are negligible. Consequently, the particles present in this medium can be considered as non-interacting entities, devoid of any mutual force exertion. The oscillations observed in the medium can no longer be adequately characterized by applying restoring forces as described by Hooke's Law. Alternatively, we can elucidate how sound waves propagate through the air medium by attributing it to fluctuations in atmospheric pressure. A sound wave can be characterized as a propagating pressure wave. Sound waves are classified as longitudinal waves since pressure variations align with the direction of wave propagation (Tomonaga, 1950).

While individual air particles do not exhibit oscillation around an equilibrium position, the propagation of a sound wave through air results in oscillations in either the air's density or pressure; in the context of Physics 7A, it was observed that air molecules exhibit rapid and random motion while concurrently exhibiting discernible average values for density and pressure. The phenomenon of sound waves in the air can be described as the fluctuation of the average particle density, leading to variations in pressure, occurring over distances significantly greater than the average inter-particle spacing. Therefore, it is possible to construct a sound wave model by solely considering the air's pressure or density (Tijdeman, 1975).

Although sound waves differ slightly from material waves, we can effectively apply most of the techniques we have already acquired to analyze and understand them. Sound waves can be understood as material waves propagating through a medium, which explains why similar techniques can be applied.

When pressure is selected as the parameter to describe sound waves, the equation representing the sound wave is transformed as follows (Buchenau et al., 1992):

$$P(x,t) - P_{atm} = P_o \sin\left(\frac{2p}{T}t - \frac{2p}{l}x + f_o\right) \tag{1.22}$$

P(x,t) is the absolute pressure of the air at a given position x and at a time t. P_{atm} is the atmospheric pressure (i.e., the equilibrium pressure). P_o is the amplitude of the pressure fluctuation (gauge pressure) from equilibrium, such that when the displacement is at a crest, the pressure is at a maximum of $P_{atm} + P_o$, and when it's at a trough, the pressure is at a minimum of $P_{atm} - P_o$ (Wittmann et al., 2003).

1.4.2. Wavefronts and Rays

The wave equation introduced in the preceding section pertains to one-dimensional waves. Nevertheless, it is commonly observed that waves exhibit propagation in either two dimensions, as seen in the surface of the water, or three dimensions, as observed in the cases of sound or light. This course will not delve into the numerical representation of higher dimensional waves. However, it is worth noting that alternative representations for multidimensional waves possess practical utility.

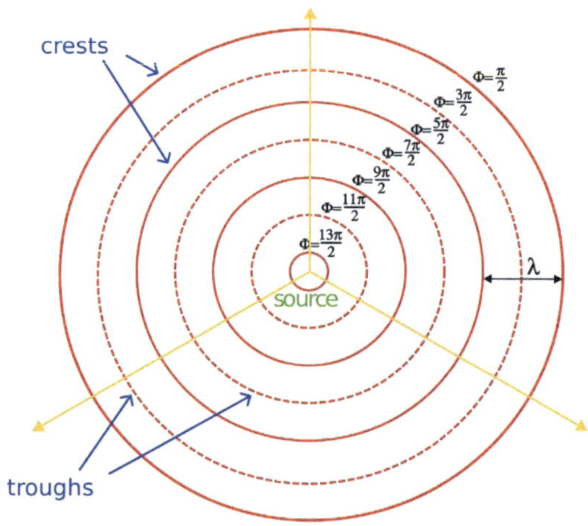

Figure 1.14. Wavefronts and rays (Source: UC Davis, Creative Commons License).

The diagram below depicts a visual representation of wavefronts in a two-dimensional space. These wavefronts are represented by concentric circles, with solid lines indicating crests and dashed lines indicating troughs. These wavefronts originate from a source situated at the center of the diagram. The propagation of wavefronts exhibits a phenomenon akin to the dispersion of ripples in a body of water as they propagate outward from the point of origin (Galindo-Rodríguez & Silva-Ortigoza, 2021).

This visual depicts the spatial arrangement of peaks and valleys at a particular moment in time. In subsequent instances, the wavefronts would progressively manifest at greater distances from the point of origin, accompanied by the generation of additional wavefronts characterized by the emergence of crests and troughs. Every wavefront offers data regarding the respective chronological order of their generation. As the wavefronts propagate radially, the wavefront farthest from the source is the initial generated, while the wavefront nearest to the source is the final one emitted. Hence, the progression of time exhibits an inward incrementality, moving from the farthest to the nearest wavefront. Assuming the location of the source is designated as $x = 0$ m and the time the first crest was generated is denoted as $t = 0$ s, it can be determined that the total phase of the furthest crest corresponds to $\pi/2$. The furthest trough was observed at a time of $T/2$ seconds, half a cycle later (Kaschwich & Gajewski, 2003).

Consequently, the total phase is greater than the initial crest by π, leading to a total phase of $3\pi/2$. The phase of each subsequent crest is incremented by a complete cycle, equivalent to an increase of 2π, relative to the phase of the preceding crest, and this pattern continues indefinitely. The wavefront representation emphasizes that the overall phase is an inherent property of a particular wave segment (Bos & Slawinski, 2010).

The wavefront representation also encompasses details regarding the spatial periodicity of the wave, as it indicates that the distance between adjacent crests or troughs corresponds to one wavelength. Nevertheless, the representation of this data as a snapshot fails to capture important temporal characteristics such as periodicity, period, or frequency (Rayces, 1964). Furthermore, due to the nature of this depiction as a "top view" of the wave propagating in a two-dimensional space, it is impossible to derive any information regarding the amplitude from this visual representation. Indeed, as will be demonstrated shortly, it is evident that amplitude ceases to remain constant in waves of higher dimensions. While this representation lacks comprehensive information about the wave, it is a valuable tool for comprehending specific wave phenomena, such as interference, which will be discussed in subsequent sections. When extended into three-dimensional space, the concentric circles depicted in Figure 1.14 would transform into concentric spherical shells (Rawlinson et al., 2008).

The arrows in the diagram offer an additional valuable means of representing waves. The arrows solely retain information on the direction of the wave. The vectors are consistently oriented perpendicular to the wavefronts and aligned in the direction of wave propagation. This representation will prove particularly valuable when examining optics as an application of diverse light properties. Numerous optical phenomena are primarily attributed to the trajectory of light rays. Therefore, the ray representation is considered the most advantageous form for that particular unit (Juárez-Reyes et al., 2015).

The majority of waves observed in nature exhibit multiple dimensions. The exclusive manifestation of a one-dimensional wave occurs solely within the constraints of a string or rope. Nevertheless, the one-dimensional wave representation is frequently deemed a suitable approximation for multi-dimensional waves located at a considerable distance from the source. Consider a hypothetical scenario in which an observer is positioned significantly from a source of waves, limiting their ability to perceive only a limited portion of the wavefronts (Vinje et al., 1996). The dashed rectangle in the Figure 1.15 represents this. From this observer's perspective, the individual wavefronts appear to exhibit a near-parallel orientation. As the distance between the observer and the source increases, the observer's ability to detect a smaller portion of the wavefronts also increases, resulting in the wavefronts appearing more parallel. When the wavefronts are parallel, it indicates that the wave is propagating perpendicular to the wavefronts. This phenomenon is depicted in the adjacent magnified image of the wavefronts located at a significant distance from the source. The perpendicular rays to the wavefronts exhibit parallelism, indicating the one-dimensional nature of the wave motion (Cocco et al., 2022).

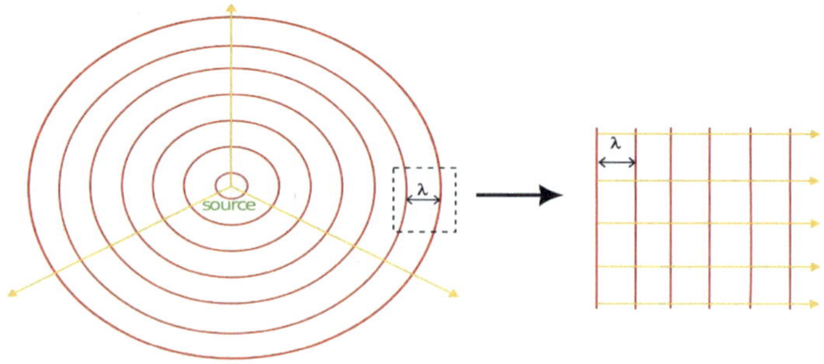

Figure 1.15. Plane waves (Source: The LibreTexts libraries, Creative Commons License).

The phenomenon of multi-dimensional waves propagating far away from the source is called plane waves. The one-dimensional wave equation is frequently employed to depict plane waves that arise from a source generating waves in multiple dimensions. The measurement is taken to quantify a single wavelength by preserving the distance between two adjacent crests (Coman & Gajewski, 2001).

1.4.3. Wave Intensity

In the preceding section, it was noted that amplitude is conserved solely in one-dimensional waves. This assumption is made under the condition that the dissipative effects of the medium can be considered negligible. In other words, the particles within the medium undergo oscillations without any frictional forces. This assumption entails that the entirety of the energy present in the wave remains confined within the wave itself, with no conversion of energy into thermal energy occurring within the medium. Nevertheless, when considering the effects of friction, the amplitude of multi-dimensional waves is no longer constant (Ramsey & Sugawara, 1997).

Let us now contemplate a wave propagating in a radial manner from a singular source in a two-dimensional space (resembling the circular disturbance created on a pond's surface due to a small stone's impact). In the medium, every position is occupied by a particle that undergoes harmonic oscillations, akin to a mass attached to a spring. As the wave propagates outward, the quantity of oscillating particles progressively augments. The uniform spacing of particles within the medium results in a consistent distribution throughout. Consequently, the circular wave encounters some particles that are directly proportional to its circumference and, consequently, directly proportional to its radius. This implies that when the wavefront radius increases by a factor of two, the number of particles oscillating in the medium also doubles. As the wave propagates, there is no dissipation of energy (Tyberg et al., 2009).

Consequently, the energy becomes distributed among more oscillatory elements when the circular region expands. The energy of each oscillator is contingent upon its amplitude of oscillation (Figure 1.16). Consequently, their amplitudes must diminish for a greater number of oscillators to possess an equivalent energy to a smaller number of oscillators (Sugawara et al., 2009).

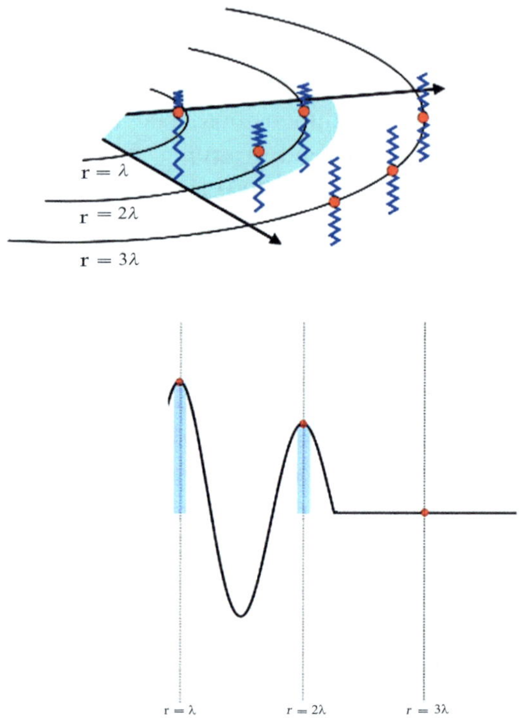

Figure 1.16. Circular wave energy conservation (Source: LibreText, Creative Commons License).

Let us examine the relationship between the oscillator's energy and its amplitude. It is important to note that the total energy of a spring-mass system can be determined by adding the kinetic energy and potential energy components together (Pavić, 1976).

$$E_{tot} = KE + PE_{sm} = \frac{1}{2}mv^2 + \frac{1}{2}kx^2 \tag{1.23}$$

Amplitude is the maximum displacement from equilibrium (x=A) when the speed of the oscillator is zero (turnaround point). This results in the following expression for the total energy:

$$E_{tot} = \frac{1}{2}kA^2 \tag{1.24}$$

Hence, the energy associated with each oscillator exhibits a direct proportionality to the square of its amplitude. This implies that when the radius of a circle is doubled, the number of oscillators between which the same energy is distributed also doubles. As a result, the square of the amplitude of each oscillation decreases by half, and the amplitude decreases by a factor of the square root of 2. When the radius is tripled, the amplitude is decreased by a factor of the square root of 3, and this pattern continues. Stated differently, the amplitude exhibits a direct proportionality to the inverse square root of the radius (Niki et al., 2002).

$$A\alpha \frac{1}{\sqrt{r}} \quad (1.25)$$

The provided diagram (Fig. 1.16) illustrates the variation in wave amplitude across a cross-sectional view as the wave transitions from a radius equivalent to one wavelength to a radius equivalent to three wavelengths.

The velocity of the wave remains constant as it propagates from the inner circle to the outer circle, implying that the energy transfer rate across each circle is uniform. The distinguishing factor between the two circles lies in the disparity of energy density encompassed within each (Langley & Bercin, 1994). In the case of the smaller circle, the energy is distributed across a smaller circumference compared to the larger circle. Consequently, the energy density diminishes as the wave propagates outward, despite the constant total energy. Power density can be defined in a similar manner, whereby the power of the wave is divided by the size of the region it traverses, which remains consistent for both rings and other areas. The term "power density" is commonly referred to as intensity. In the context of a two-dimensional wave, the quantity mentioned above refers to the ratio between the power of the wave and the circumference of the circle that it traverses (Sun et al., 2000).

$$I_{2d}(r) = \frac{P}{2pr} \quad (1.26)$$

Hence, the magnitude of a two-dimensional wave propagating radially from a focal point exhibits an inverse relationship with the distance from the central origin. It is observed that the intensity exhibits a direct proportionality to the square of the amplitude (Fujimoto et al., 2004).

$$A\alpha \frac{1}{\sqrt{r}} \Rightarrow I\alpha A^2 \quad (1.27)$$

The observation reveals that the relationship between intensity and square amplitude also holds true in the context of one-dimensional systems. In the context of a one-dimensional wave, it can be observed that the energy density remains constant. This is because the energy is transferred exclusively between adjacent oscillators, with each oscillator passing its energy to its neighboring oscillator. Hence, it can be observed that the power density, also known as intensity, remains constant. This observation aligns with our existing knowledge that the amplitude of a wave in one dimension remains unchanged (Kingsley et al., 2017).

Three-dimensional waves with central sources, such as sound and light, are significantly more prevalent in our research. In these instances, the calculation of power density necessitates the division by the surface area of a sphere, as opposed to a circle. In this particular instance, the magnitude of the wave is measured in watts per square meter, whereas the magnitude of the two-dimensional wave was measured in watts per meter. Thus, we can express the relationship as follows (Kinnunen et al., 2019):

$$I_{3d}(r) = \frac{P}{4\rho r^2} \tag{1.28}$$

Once again, we find the same relationship between intensity and amplitude. The same mechanism works: as the wave moves outward from a central point, the number of oscillators on each spherical surface is proportional to the surface area. Doubling the radius of a spherical surface quadruples the surface area, so the number of oscillators grows with the square of the radius. This means that the energy per oscillator drops with the square of the radius, and the amplitude is inversely proportional to the radius:

$$A\alpha \frac{1}{r} \Rightarrow I\alpha A^2 \tag{1.29}$$

The relationship between intensity and amplitude is universally applicable to all waves, and it is a concept that will be consistently considered in the subsequent sections (Chrysohoou et al., 2015).

1.5. DOPPLER EFFECT

The Doppler Effect is an intriguing phenomenon in physics that manifests when a relative motion exists between a wave source and an observing entity. This statement elucidates the phenomenon of frequency alteration, which consequently affects the perceived pitch of waves, including sound and light waves, when there is relative motion between the source and observer. The phenomenon was initially elucidated by Christian Doppler in 1842 and has subsequently been applied in practical contexts across diverse disciplines, encompassing astronomy, meteorology, and commonplace scenarios such as perceiving the sound of sirens emitted by mobile vehicles or adjusting the pitch of musical instruments (Reed et al., 2003).

The Doppler Effect can be comprehended by contemplating a locomotive in motion emitting a whistling sound. When the train is at rest, the sound waves emitted by the whistle propagate evenly in all directions, resulting in a consistent frequency perceived by an observer on the platform. Nevertheless, if the train is approaching the observer, the sound waves located ahead of the train undergo compression, increasing frequency, thereby causing the observer to perceive a higher pitch. In contrast, as the train recedes from the observer, the sound waves undergo elongation, resulting in a decreased frequency and a lower pitch perceived by the listener. The fundamental characteristic of the Doppler Effect is the alteration in frequency (Voti & Bertolotti, 2021).

The Doppler Effect's mathematical formulation applies to sound and light waves. The formula incorporates the frequency of the source, the speed of sound or light in the given medium, and the velocities of both the source and the observer concerning one another. Spectroscopy in astronomy is significant, as astronomers employ this technique to ascertain the movement of celestial entities by examining the redshift or blueshift observed in their spectral lines. Doppler radar is employed in meteorology to detect precipitation particles' movement, thereby serving as a vital tool in weather forecasting. Additionally, it assumes a critical function in medical imaging, facilitating the visualization of blood flow and aiding in diagnosing cardiovascular ailments (Luo et al., 2020).

Although widely utilized and deemed valuable, the Doppler Effect is subject to certain limitations and considerations. The assumption is that sound waves propagate through a medium at rest while light waves propagate through a vacuum. Additionally, the observed phenomenon is contingent upon the relative motion between the source and observer along the line of sight. Moreover, the consideration of relativistic effects becomes necessary when addressing the behavior of entities in motion at significant velocities, necessitating the utilization of specialized equations derived from the theory of special relativity to incorporate the influence of the Doppler Effect precisely (Raman & Nagendra Nath, 1936).

In summary, the Doppler Effect is a fundamental and practical principle in physics with wide-ranging implications in numerous scientific, technological, and everyday contexts. The comprehension of this subject matter has significantly enhanced our understanding of the cosmos, made notable contributions to the field of weather forecasting, facilitated advancements in medical diagnostics, and improved communication systems. The Doppler Effect remains a pertinent and influential phenomenon in various domains of human existence, and it continues to be a prominent subject of investigation within the realm of contemporary physics (Bolotovskiĭ & Ginzburg, 1972).

1.5.1. Stationary Source and Observer

Thus far, in examining wave characteristics such as wavelength, frequency, and speed, we have operated under the assumption that both the source generating the wave and the observer detecting and measuring its properties are in a state of rest (Figure 1.17). The animation below illustrates a scenario in which the central dot, denoted by the color red, serves as the source that emits a crest each time it flashes. The depicted wave representation illustrates a two-dimensional wave, wherein the circular shapes symbolize the crests that originate from a source, akin to the ripples observed in a body of water. The spacing between two adjacent circles corresponds to the wavelength of the wave. Given that the visual depiction is animated, it is also possible to ascertain the period by measuring the frequency of occurrences in which the red dot flashes (Martirosyan et al., 1984).

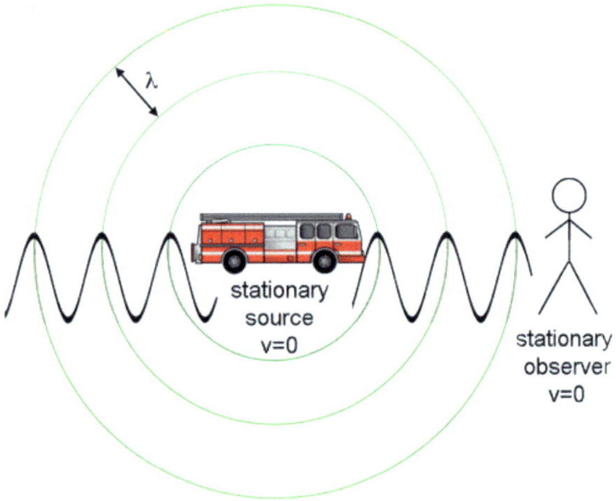

Figure 1.17. Source and observer are stationary (Source: Source: NASA's Imagine the Universe), Creative Commons License)

A blue dot represents the observer located to the left of the source. This dot exhibits a flashing behavior each time a crest traverses its position. The concept of an observer holds particular relevance in the context of sound waves, as they are directly perceived and interpreted in our everyday experiences. When the peaks of the waves apply force to the auditory organs, the tympanic membranes undergo oscillatory motion. The human brain perceives the magnitude of these oscillations as the auditory sensation of volume, while the rate of oscillations is perceived as the auditory sensation of pitch.

When the source and observer are both at rest, the frequency, wavelength (the distance between two consecutive crests), and wave speed will be identical. The subsequent inquiry we aim to pose pertains to the consequences that arise when there is involvement of motion, either by the source, the observer, or both (Alexander, 1997).

1.5.2. Moving Source

The animation below depicts the source's movement toward an observer at rest (Figure 1.18). This source can be likened to a rapidly moving ambulance emitting a distinct high-pitched sound. Upon careful observation of the animation, it becomes apparent that there exists a discrepancy between the emission frequency of the source and the reception frequency by the observer. Specifically, the observer, represented by the blue dot, exhibits a higher flash frequency than the source, represented by the red dot. In what manner can the observer measure a frequency that deviates from the emitted frequency of the source (Davys et al., 1985)?

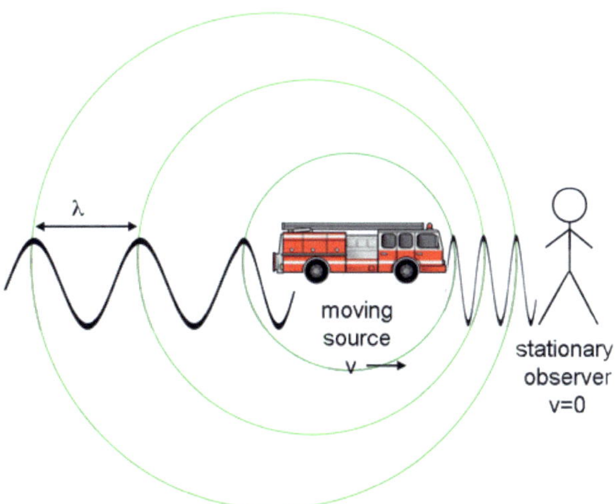

Figure 1.18. Source moving toward stationary observer (Source: NASA's Imagine the Universe), Creative Commons License).

Let us endeavor to comprehend this phenomenon from the observer's vantage point. The animation indicates a noticeable disparity in crest density, with a higher concentration observed in the vicinity of the advancing source and a lower concentration observed in the region behind the source. Therefore, the observer before the moving source perceives a shorter wavelength. When the source remains at rest, the separation between two successive crests is equivalent to the source wavelength, denoted as λs, representing the wavelength determined by the source itself. Nevertheless, in the case of a mobile source, it is observed that after emitting a crest, the source undergoes a specific displacement before emitting the subsequent crest. Consequently, as it emits subsequent crests, the wave progressively approaches the crest preceding it, reducing the distance separating adjacent crests. Similarly, as it propagates, it progressively moves farther from the preceding crest, resulting in increased separation between successive crests originating from the same source (Schulkes et al., 1987).

Determining the new distance between the crests can be accomplished by computing the displacement of the source between successive emissions of crests. The duration required for the emission of two adjacent crests is referred to as the source period, Ts, measured in terms of the source's period. The velocity at which the source is in motion is known as the source speed, vs., resulting in a displacement of d = vsTs. Hence, the observed wavelength (λo) in the presence of a moving source can be determined by subtracting the distance the source covers in one period from the source wavelength (Hirasawa & Stauder, 1965).

$$\lambda_o = \lambda_s - v_s T_s \tag{1.30}$$

From this, one can ascertain the frequency at which the observer makes measurements. The relationships $\lambda = v/f$ and $T = 1/f$ can be employed to express equation 1.30 regarding frequencies. The symbol "v" will be utilized without any subscript to denote sound velocity. It is important to note that in this particular scenario, the speed of sound remains constant for both the source and the observer. This constancy is attributed to the properties of the medium through which sound travels. It should be mentioned that the observer, in this case, is stationary. However, it will be demonstrated later how the speed of sound varies when observed from a moving perspective. By substituting frequency for wavelength in Eq. (1.30), we obtain the following (Yeung & Nguyen, 1999):

$$\frac{v}{f_o} = \frac{v}{f_s} - \frac{v_s}{f_s} \tag{1.31}$$

By rearranging the equation, it is possible to articulate the observed frequency in relation to the source frequency, employing the same methodology as previously described.

$$f_o = \frac{v}{v - v_s} f_s \tag{1.32}$$

It can be observed from the equation mentioned above that the frequency observed is greater than the frequency of the source when the source moves in the direction of the observer. This is since the numerator, v, is larger than the denominator, v−vs. This observation aligns with the blue observer dot exhibiting a higher frequency of flashes compared to the red source dot (Zolotukhina, 1975).

When the observer is positioned behind the moving source, the animation demonstrates that the crests of the waves exhibit an increased separation as the source moves away. In this scenario, the source undergoes a displacement of vsTs away from a previously emitted crest before emitting the subsequent crest. Therefore, the wavelength that is observed is:

$$\lambda_o = \lambda_s + v_s T_s \tag{1.33}$$

Resulting in the following expression for the observed frequency:

$$f_o = \frac{v}{v + v_s} f_s \tag{1.34}$$

In the given equation, it can be observed that the frequency observed is lower than the frequency of the source. This is since the numerator, v, is smaller than the denominator, v+vs. Due to the fixed nature of the speed of sound, a shorter wavelength preceding an approaching source leads to an increase in frequency (Miles, 1971). In contrast, a longer wavelength trailing a receding source results in a decrease in frequency. In future instances when an ambulance is rapidly passing by, it is advisable to direct your attention towards the auditory characteristics of the siren as it approaches and subsequently moves away from your location. This phenomenon is commonly referred to as the Doppler effect.

Combining the two equations can derive an expression for the Doppler effect in the context of a moving source.

$$f_o = \frac{v}{v \pm v_s} f_s \qquad (1.35)$$

where the sign is a plus for a source moving away from the observer and a minus for a source moving toward the observer (Svirkunov & Kalashnik, 2014).

1.5.3. Moving Observer

The observer moves toward or away from a stationary source, not the source itself. Figure 1.16 shows that the wave crests' spatial spacing stays constant in front of and behind the stationary source. In this case, the wavelength is constant. The distance between wave crests would be the same for a moving observer. In Figure 1.16, the flashes should grow as the blue dot approaches the source. The viewer would experience crests more often than while immobile. Thus, the frequency rises. It is unknown how these phenomena may occur if the wavelength remains constant and the medium does not change (Thevenot & Kraus, 1995).

Analogize this. Imagine driving near a train on parallel lines next to the road. The train represents the dynamic wave, while the mobile car houses the observer. Train inter-car distances are uniform, implying constant wavelengths. Moving in the opposite direction of the train increases the rate at which one sees each new vehicle. Train velocity is higher than the perceived velocity at rest. The observer's perception of the train's speed is increased despite its constant velocity. If an individual traveled in the same direction as the train, the train would appear to slow down or even stop if their speeds were the same. Relative motion is discussed, where object velocity depends on the reference frame. When watched from a moving vehicle, a train's measured speed will differ from a stationary observer.

Thus, sound speed increases when the observer moves closer to the wave-emitting source. v + vo, where v is the reference frame's sound speed, and vo is the observer's speed, determines this occurrence. Thus, the seen frequency equals the source's wavelength divided by the observer's speed of sound (Spengler et al., 2000).

$$f_o = \frac{v + v_o}{l} = \frac{v + v_o}{v} f_s \qquad (1.36)$$

where wavelength was v/fs. If the observer is traveling away from the source or driving in the same direction as the moving train, the speed recorded by the observer lowers, and the speed of sound becomes v−vo. The measured frequency for an observer traveling away from the stationary source is: Eq. (1.37) applies.

$$f_o = \frac{v - v_o}{v} f_s \qquad (1.37)$$

Combing the two equations, the Doppler effect for a moving observer and a stationary source is:

$$f_o = \frac{v \pm v_o}{v} f_s \qquad (1.38)$$

The observed frequency is higher than the source one when the numerator is bigger than the denominator when the sign is positive. This is consistent with the observation of the blue dot flashing more frequently if it was to move toward the source. On the contrary, we choose a negative sign when the observer moves away from the source, resulting in a lower observed frequency (Rousseaux et al., 2008).

1.5.4. Combined Doppler Effect

Let's explore the Doppler effect when both source and observer move for completeness. When the observer and the source are moving, the observer senses a separate sound speed and wavelength due to their relative motion. Thus, the observer's velocity of sound is v±vo, and the wavelength is o. Thus, the frequency expression (Nezlin, 1976):

$$f_o = \frac{v \pm v_o}{l_o} \qquad (1.39)$$

The observer's wavelength matches a moving source's. The total Doppler effect is expressed using Eq. (1.36):

$$f_o = \frac{v \pm v_o}{v \pm v_s} f_s \qquad (1.40)$$

It can be observed from the abovementioned equation that in the case where the observer is at rest, vo = 0, Eq. (1.35) is obtained, corresponding to the scenario of a stationary observer and a source in motion. When the velocity of the source, vs., is equal to zero, we can derive Eq. (1.38), which describes the scenario of a stationary source and a moving observer. In the absence of relative motion between two objects, the Doppler effect does not occur, leading to the equality of the observed frequency (fo) and the source frequency (fs) (Censor & Newhouse, 1988).

Since there are many variables while solving Doppler effect issues, below is a list of all the terms in Eq. (1.40):

- The observer's measured frequency is fo.
- Source frequency is "fs."
- Wave speed, v, is the velocity at which a wave propagates across a medium. Sound waves' airspeed is often described as thus. The observer's speed, vo, should be written with a plus sign in the numerator while traveling toward the source and a negative sign when going away.

When the source approaches the observer, vs. is negative in the denominator, and when it is retreating, it is positive (Ziemkiewicz & Zielińska-Raczyńska, 2015).

Other waves also exhibit the Doppler effect. In 1842, Austrian mathematician and scientist Christian Doppler discovered the Doppler effect by observing light from distant celestial bodies. The Doppler effect showed that light from distant celestial bodies had a redshift, indicating longer wavelengths than expected. This led to the idea of an expanding universe. According to the expanding universe theory, our nearby celestial bodies always recede or move. Due to electromagnetic waves' transmission at the speed of light, Eq. (1.40)'s Doppler effect does not apply to them. Relativistic effects must be included in Doppler effect calculations.

Ultrasound technology uses the Doppler effect. Police radar measures vehicle speeds using the Doppler effect. The radar system detects a wave reflected by the mobile vehicle. However, radars emit radio waves, which need relativistic Doppler equations (Qiu et al., 2022).

1.6. SUPERPOSITION AND INTERFERENCE

Wave dynamics, superposition, and interference explain how waves interact and generate complicated patterns. Whether studying sound, water, light, or electromagnetic waves, the ideas covered here are crucial to understanding wave behavior and phenomena. This explanation explores superposition and interference and their relevance across fields (Kolokolov & Skrotskiĭ, 1992).

The superposition principle states that when several waves cross a spatial domain, they can coexist and propagate separately without interference. Instead of harmful interference, the waves exist concurrently and combine to create a new wave with the summing of their amplitudes at any given point. The resulting wave is the superposition of the original waves (Van Kruining et al., 2018).

In mathematical terms, wave superposition occurs when two waves, A and B, with respective amplitudes A1 and A2 and frequencies f1 and f2, are combined. The resulting wave, C, at any specific time and location can be determined by the equation C = A1 * sin(2f1t) + A2 * sin(2f2t). This applies to electromagnetic and mechanical waves like light and sound (Noel & Stroud, 1996).

The superposition principle explains interference, which occurs when two or more waves combine to enhance or weaken their amplitudes. Where waves reinforce one other, constructive interference increases amplitude and intensity. However, destructive interference occurs in locations where waves cancel each other, decreasing amplitude and sometimes annihilating them.

Light waves are best for interferometry. The famous double-slit experiment illuminates two closely spaced slits. A screen behind the slits records the light pattern. Interference fringes form when light waves pass through apertures and interfere. Light waves collide and merge, causing constructive and destructive interference (Kovachy et al., 2015).

Interference can be classified into two main types:

- ***Constructive Interference*:** In constructive interference, the crest of one wave aligns with the crest of another wave, or the trough of one wave aligns with the trough of another wave. When this alignment occurs, the individual amplitudes of the waves add up, resulting in a larger combined amplitude and increased intensity. Constructive interference leads to the formation of bright regions in the interference pattern (Kawase, 1996).
- ***Destructive Interference*:** In contrast, destructive interference occurs when the crest of one wave aligns with the trough of another wave. This alignment causes the individual amplitudes to cancel each other out, reducing the combined amplitude and diminished intensity. Destructive interference leads to the formation of dark regions in the interference pattern.

Interference affects several fields. Interference is used in diffraction gratings, Fabry-Perot interferometers, and anti-reflective coatings. Audio engineers use interference to create flanging and chorus. Interference can degrade radio signal reception. Musical instruments also suffer from interference. In conclusion, superposition and interference control wave behavior and provide insights into many natural and artificial systems. The superposition principle allows waves to coexist and interact, creating various wave shapes. Superposition creates positive and negative interference zones. These zones shape light patterns, musical instrument sounds, and interference fringes. These principles are crucial in optics, audio engineering, telecommunications, and cosmology. Superposition and interference continue to illuminate nature's mysteries as we study waves (Polifke et al., 2001).

1.6.1. Overview of Superposition

We have investigated how a unique source generates a wave pulse, periodic wave, or harmonic wave. Understanding the source's motion and energy input helps categorize a disturbance as it propagates at the medium's speed. We also considered source and observer motion. All previous discussions assumed a single-wave source.

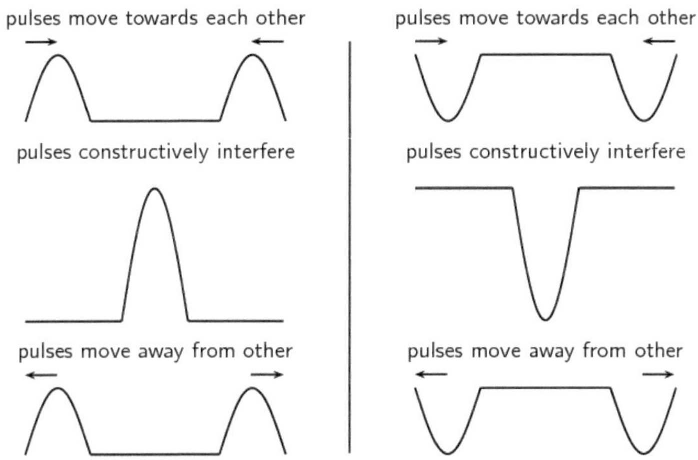

Figure 1.19. Superposition of two pulses: constructive interference (Source: Siyavula, Creative Commons License).

We are presently studying the spatial intersection of two waves with comparable features (Figure 1.19). Two people holding a rope together illustrate this. A posterior oscillation sends a wave toward a partner. Your partner's oscillations will reach you. Another example is dropping two stones into the water. Wave collision effects? The waves will converge with time, but it is unclear how to calculate the displacement from equilibrium using math. Superposition is the combination of waves to create an effect. Wave amplitudes add when they cross. The animation shows two wave pulses converging during the overlapping period. After overlap, the pulses maintain their initial direction and amplitude (Di Benedetto et al., 2011).

For two one-dimensional waves of the same type, superposition can be written as:

$$y_{total}(x,t) = y_1(x,t)+y_2(x,t) \quad (1.41)$$

The same principles apply to continuous harmonic waves, which specifically can be represented as:

$$y_{total} = A_1 \sin\Phi_1 + A_2 \sin\Phi_2 \quad (1.42)$$

Let us look at an example of two waves with the same wavelength and amplitude meeting at the same location in space graphically. Since these waves change with time, we can only see a snapshot of this occurrence, as shown in Figure 1.20. Although the two waves, y1, and y2, overlap in the same location, the wave is plotted separately for clarity (Deckers et al., 2014).

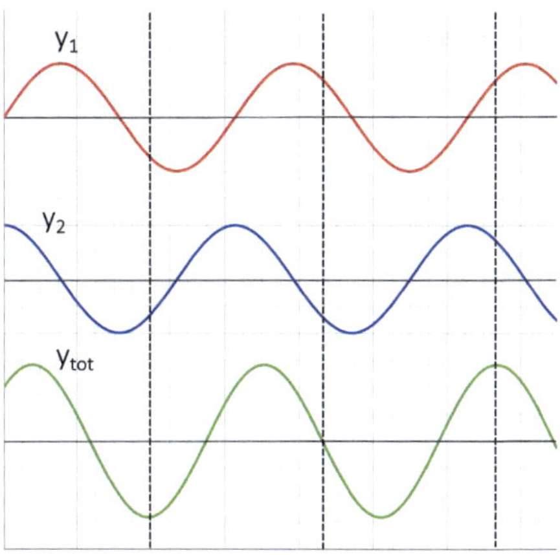

Figure 1.20. Superposition (Source: Dina Zhabinskaya, Creative Commons License).

The plots show two waveforms, y1, and y2, that converge at a specified time and place. The lower figure, ytot, shows the cumulative total of the two waves. Add the displacements of each wave at each position to total them. Vertical dashed lines indicate several displacement addition places. The midpoint is zero since y1 is positive and y2 is negative. Since the crests and troughs never coincide, the summation of the separate waves' amplitudes does not equal the overall waves. Both waves have an amplitude of two grid lines, but the combined wave approaches three grid lines. As shown later, this assumption is not always valid. In severe cases, the resulting amplitude may be zero (Kurner et al., 1993).

1.6.2. Special Cases of Interference

Wave phenomena use superposition primarily through interference. Two ways to mix sinusoidal functions are very important. The left panel shows the exact overlap of the two waves. The superimposed wave's amplitude equals the total of each location's displacements since they are uniform. It's called constructive interference. The net displacement is zero when one wave rises, and another falls. Destructive interference occurs when two waves are half a cycle apart. As seen in the right panel of the figure below, the two waves' superposition provides a zero net value at all sites, nullifying the waves. Figure 1.20 shows partial interference, which is neither constructive nor destructive. "Partial interference" is too broad because it includes positive and harmful interference (McCumber, 1968).

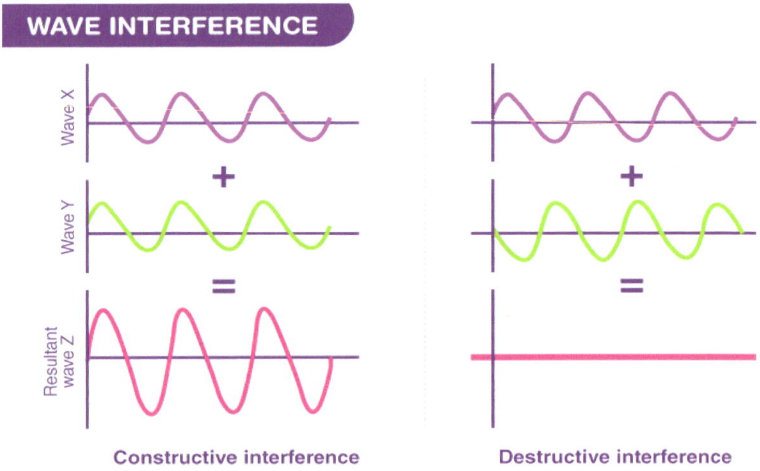

Figure 1.21. Constructive and destructive interference (Source: BYJU's, Creative Commons License).

The alignment of the two waves determines interference, as seen in the pictures. When waves are in phase, their amplitudes are the total of their crests and troughs. Thus, the sine function is preserved since the waves' phases match or deviate by an integer multiple of 2. In Eq. (1.42), the two phases differ only by a multiplication factor 2n. Eq. (1.42) for constructive interference is:

$$y_{total} = A\sin\Phi + A\sin(\Phi + 2\pi n) = 2A\sin\Phi \quad (1.43)$$

The combined wave on the left plot of Figure 1.20 shows that 2A is double the original amplitude.

In contrast, wave cancellation occurs when one wave's peak coincides with another wave's trough, resulting in an odd multiple. Figure 1.20 shows the illustration on the right. Eq. (1.42) changes to the following expression:

$$y_{total} = A\sin\Phi + A\sin(\Phi + 2\pi(n+1/2)) = A\sin\Phi - A\sin\Phi = 0 \quad (1.44)$$

This uses the sine function's feature that altering its phase by half a cycle negates it.

These examples show that the type of interference detected depends on the phase difference of the two waves. Interference between identical-frequency waves will be our focus. Wavelengths must be equal to interference in a medium. The position is usually measured from the source in interference concerns. To explain, x positively associates with time, suggesting a rightward wave. As previously stated, the spatial term in the equation for the overall phase will be denoted by the negative sign (Kiselev et al., 1975).

$$\Phi = \frac{2p}{T}t - \frac{2p}{l}x + f_o \quad (1.45)$$

Three things can affect the phase difference between two waves. Thus, any of the three elements can reveal the phase difference between two waveforms. The time-dependent component, 2t/T, may vary if equal-frequency sources were engaged at various times. For instance, two speakers engaged simultaneously have equal phase constants. These speakers emit waveform crests and troughs simultaneously. If one activated one speaker half a period after the other, one would create a crest and the other a trough, putting them out of phase by half a cycle. However, this circumstance is similar to activating two speakers simultaneously with a phase constant differential of half a cycle, written as o =. Thus, phase constant variation may reliably accommodate any temporal discrepancy in wave creation. To simplify the study, we will assume that the time difference component, t, is zero by regularly synchronizing both sources' time. The entire phase difference becomes:

$$\Delta\Phi = -\frac{2p}{l}\Delta x + \Delta\phi_o \quad (1.46)$$

The equation's first term depends on each source's spatial arrangement relative to the interference point (which will be explained later). In contrast, the second term represents the phase constant difference between the two sources.

Interference causes phase mismatch under certain conditions:

- When the entire phase difference is an even multiple, n is an integer (n = 0, ±1, ±2,...).
- Destructive interference occurs when the phase difference between two waves is an odd multiple of, represented as n = 2n+1/2.
- Partial interference occurs when the overall phase difference between two waves does not match the abovementioned parameters (Roshchupkin & Lysenko, 1999).

1.6.3. Interference of Sound Waves

Sound wave interference will be examined in this study. Combining two waves yields the largest amplitude in constructive interference. This is the highest audible intensity produced by two sources emitting sound simultaneously. Destructive interference reduces sound amplitude to zero, eliminating auditory awareness. Our study will initially be confined to sources with identical frequencies.

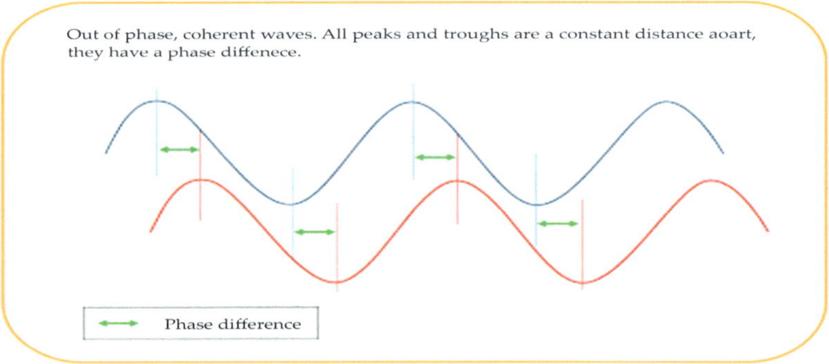

Figure 1.22. Phase constant difference (Source: Google Sites, Creative Commons License)

Eq. (1.46) shows that phase difference is affected by spatial location and phase constant differences. Examine the effects of these two words. The observer is to the right of two speakers in simple one-dimensional settings.

As shown in the picture, we want to measure the interference along the dashed vertical line right of the two sources. For clarity, the graphic shows the two sources vertically separated. Since the sources are side-by-side, their waves overlap. Waves from both sources are balanced in the image. A quarter of a cycle earlier, source 1 crest and source 2 trough. This suggests a phase constant deviation of. Visually, the crests and troughs match, causing destructive interference (Daigle et al., 1978).

Since the two sources are at the same location, Δx = 0. The difference in their phase constants is π. This results in the following expression for the total phase difference:

$$\Delta\Phi = -\frac{2p}{l}\cdot 0 + p = p \qquad (1.47)$$

This confirms destructive interference, which is observed in Figure 1.21. Note this scenario could also be interpreted as two sources with equal phase constants being turned on half a period after one another. This would mean a difference in time between the two sources, but as we already discussed, any difference in time can always be interpreted as a difference in a phase constant, such as we treated this problem here.

Figure 1.22 shows another example where the phase difference is caused by the difference in the position of the two sources. Source 2 is behind source 1 by a distance of one wavelength. The two sources have the same phase constant since both produce identical waves in this snapshot. The only difference between the two sources is their relative positions (Wei et al., 2021).

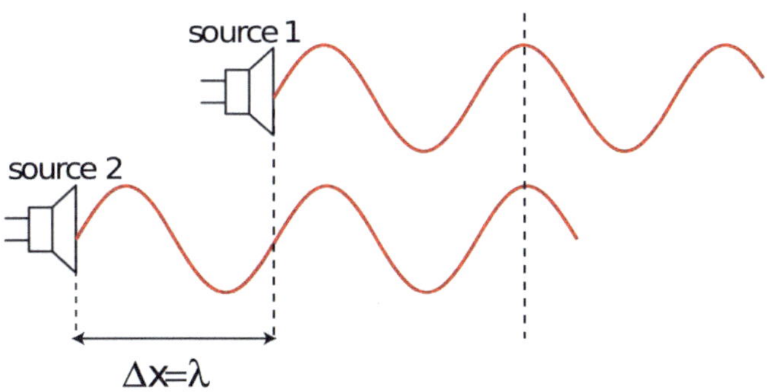

Figure 1.23. Path length difference for constructive interference (Source: UC Davis Library, Creative Commons License).

This situation is important because the wave from source 2 must travel farther to reach the observer than from source 1, even if both waves are in phase with identical phase constants. x is the route length difference between these two spatial places. The path length difference is one wavelength, written as x = x2 − x1 =. At the last vertical dashed line, an observer right of both speakers makes this conclusion. The route length difference is comparable to two speakers at the same place. This causes constructive interference. Figure 1.21's cumulative phase difference may be calculated as follows:

$$\Delta\Phi = -\frac{2p}{l}\lambda + 0 = -2\pi \quad (1.48)$$

Figure 1.23 shows a similar situation, except now source 2 is half a wavelength behind source 1, such that Δx = x2 −x1 = λ/2 as can be seen in the figure, in this case, crests and lining up with troughs, which will be resulting cancellation of amplitudes (Überall, 1998).

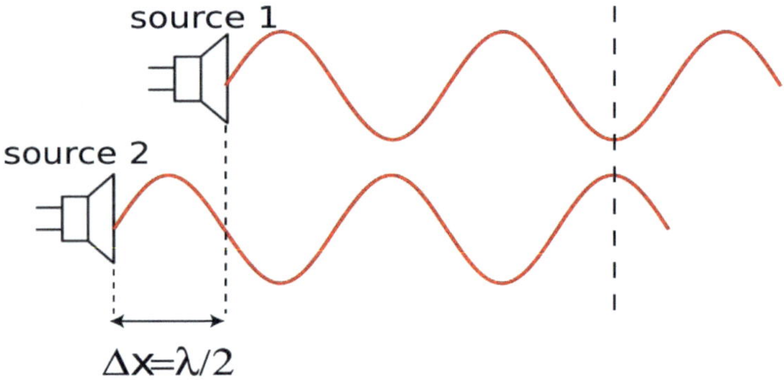

Figure 1.24. Path length difference for destructive interference (Source: UC Davis Library, Creative Commons License)

Mathematically, plugging in for the path length difference, we obtain the result for destructive interference:

$$\Delta\Phi = -\frac{2p}{l}\frac{l}{2} + 0 = -\pi \quad (1.49)$$

The right speakers' positions are consistently different. We depicted waves just propagating right of the speakers. These waves are created in a three-dimensional area around the speakers. Due to source 2's closeness, x2 – x1 would be inverted on the left speaker. The route length difference concepts still apply when measuring interference at a place where the speakers and observer are not aligned. This figure illustrates route length disparity (Figure 1.24) (Waterhouse, 1955).

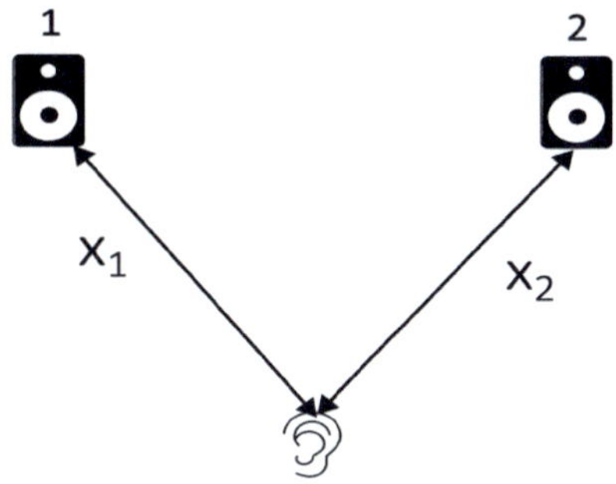

Figure 1.25. Path length difference (Source: The California State University, Creative Commons License)

The observer is x1 from speaker 1 and x2 from speaker 2. The path length difference is x = x2 − x1 (1.50) (Figure 1.25).

The route length difference between sites equidistant from each speaker is insignificant when quantifying interference. Calculating the fractional value of each source's wavelength-related distance is essential. Calculating the overall phase difference after labeling sources "1" and "2" requires consistency. Subtract speaker 1's phase constant from speaker 2's.

The last figure shows how phase constant and route length changes interact (Figure 1.26).

Source 2 is half a wavelength behind source 1 in phase constant.

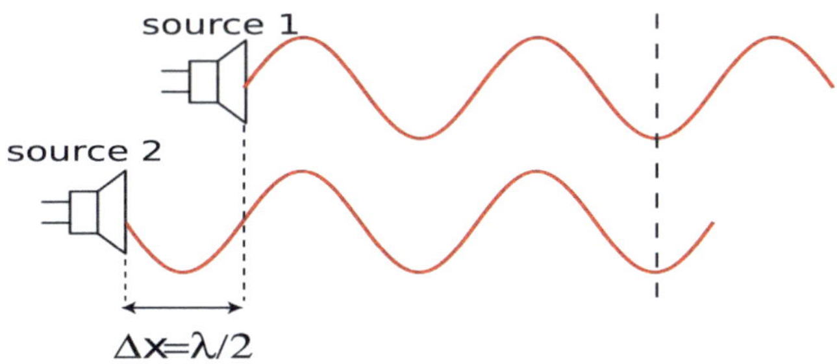

Figure 1.26. Path length difference and phase constant difference (Source: UC Davis Library, Creative Commons License)

Plugging this scenario into the equation for the total phase difference, we obtain the following:

$$\Delta\Phi = -\frac{2p}{l}\frac{l}{2} + \pi = 0 \qquad (1.51)$$

The graphic shows constructive interference by aligning wave crests and troughs. This illustrates the importance of considering both impacts. Figure 1.21 shows destructive interference from a half-cycle phase difference. Figure 1.23 shows destructive interference occurs when path length differences are half a wavelength. In this case, the effects that separately cause destructive interference produce constructive interference (Kim et al., 2011).

1.6.4. Wavefront Representation of Interference

The wavefront representation can be employed as another method for illustrating wave interference. The diagram below illustrates two distinct sources producing waves with identical wavelengths and, consequently, equivalent frequencies. The crests, represented by solid lines, and troughs, represented by dashed lines, emitted by source A exhibit a blue coloration, whereas the crests and troughs emitted by source B display a red hue. The proximity of the two sources allows for the convergence and superposition of their wavefronts at a common spatial location (Berry et al., 1980).

Did You Know?

In interference of waves, the wavefronts represent regions of constructive or destructive interference. Constructive interference occurs when wave crests coincide, resulting in increased amplitude, while destructive interference occurs when wave crests and troughs coincide, resulting in reduced amplitude.

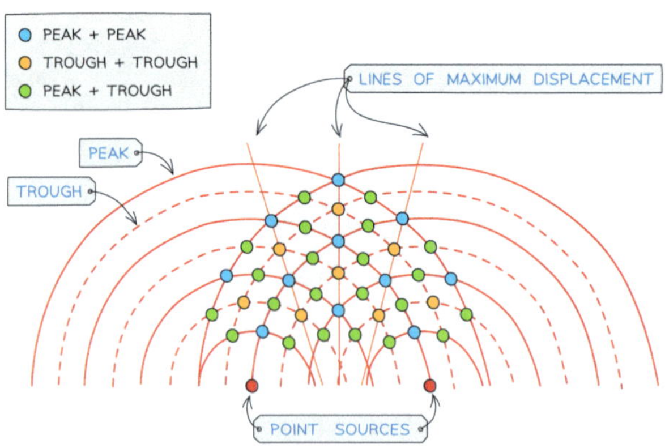

Figure 1.27. Interference between two wavefronts (Source: Save my Exams, Creative Commons License).

We know constructive interference happens when two waves intersect with two crests or two troughs. The graphic shows several places. The middle line between the sources shows constructive interference. The figure has constructive interference at overlapping crests and troughs (Figure 1.27). Overlapping crests and troughs limit constructive interference to specified places. Both sources create crests and troughs, indicating phase coherence. Instead, replacing the crests in one source with troughs and the troughs with crests would show two sources half a cycle out of phase. Due to equal source distances, the route length difference along the center axis is zero. Constructive interference occurs because the cumulative phase difference along the center axis is always zero (Felsen, 1984).

At precise places, when a crest from one source coincides with a trough from another source, destructive interference reduces or cancels amplitudes. Above is a snapshot. Future wavefronts will have more constructive and destructive interference spots. The wavefront representation cannot fully explain interference since it only shows crests and troughs at one time. It's a good reference. Predicting interference sites requires careful path length calculations, unlike the center line, which is symmetrical.

A and B are creating acoustic waves. The graphic shows two source A crests (solid red circles) and two source B troughs (dashed blue circles). All elements are proportionally represented (Sweeney et al., 2014).

1.7. BEATS

Beats are fascinating because two waves with slightly different frequencies interact. This interaction creates a pulsatile audio or visual pattern by cyclically fluctuating magnitude. Music, acoustics, and spectroscopy use sound and light wave beats. In a medium, two waves with marginal frequency differences superpose and interfere. Waves with close frequency but different crests and troughs align periodically, causing constructive interference and enhanced amplitude at select spots. Interference and amplitude loss occur when one wave's peaks overlap with another's troughs. Constructive and negative interference causes beats (Tucker, 1950).

The difference in initial wave frequencies determines the beat frequency (f_beat). The beat frequency, f_beat, is the absolute value of the difference between the two wave frequencies, f1 and f2. The beat frequency controls the wave's amplitude fluctuation. Sound waves contain auditory rhythms. When performed simultaneously, two musical tones or sound waves with modest frequency variances produce a pulsing sound. Tuning musical instruments uses these phenomena. Musicians use this approach by altering the tension of one string to match the frequency of another, relying on the auditory impression of beats when both strings are played.

Lightwave beats are rarer in daily life. Light beats occur when two coherent light sources with slightly different frequencies cause spatial interference. This interference pattern produces alternating brilliance and obscurity. Light beats are crucial to spectroscopic methods that accurately determine atomic and molecular frequencies. A, f_beat, and t indicate the beat's amplitude, frequency, and time. The first waves' constructive and destructive interference causes a cyclical magnitude variation (Misquitta et al., 1999).

The following are the applications of beats:

- ***Tuning Musical Instruments*:** Beats are used in tuning musical instruments, such as pianos and guitars. Musicians adjust the tension or length of strings until the beats between the tuned and reference strings disappear, indicating that both strings are in tune.
- ***Acoustics and Sound Engineering*:** Understanding beats is crucial in acoustics and sound engineering for designing sound systems, eliminating interference, and creating audio effects.
- ***Spectroscopy*:** In spectroscopy, beats are used in various techniques to measure precise frequencies and energy levels of atoms and molecules (Hosokawa et al., 2020).

Beats only occur when the interacting waves have close but not identical frequencies. The interference pattern from large frequency differences is too fast for humans to recognize as beats.

Beats, a fascinating phenomenon, emerge when two waves with slightly differing frequencies combine to generate a throbbing pattern with amplitude fluctuations. Music, acoustics, and spectroscopy use these phenomena in sound and light waves. Beats shed light on wave processes and interference, advancing several scientific and technical fields (Hasselmann et al., 1963).

Activity 1.1

You are asked to explain the concept of the Doppler Effect to a group of students in a science fair. How would you demonstrate and explain the Doppler Effect using simple and engaging examples? Additionally, what real-life applications can you share with the students to illustrate the practical significance of the Doppler Effect?

SUMMARY

This chapter examines wave fundamentals, including their features, behaviors, and interrelationships. Waves are introduced with a detailed discussion of their properties and use across disciplines. The next section defines waves, pulse waves, and periodic waves. Waves' dimensionality and polarization allow them to propagate in different directions. The chapter covers wave speed, representation techniques, waveforms, and basic harmonic motion, providing useful insights into wave visualization and mathematical description. The harmonic wave equation explains wave periodic motion and sinusoidal representation. This study discusses multi-dimensional waves, their capacity to propagate in numerous directions, and their ramifications. Sound waves, the most common waveform, are then highlighted. Wavefronts and rays explain sound wave transmission in space. To understand wave signal magnitude and energy, wave intensity is analyzed. The chapter also explains the Doppler Effect, which occurs when a wave source and observer move relative. This debate covers stationary, moving, and moving observer circumstances. The combined Doppler Effect is explained. This chapter concludes with superposition and interference. Superposition explains how waves coexist and combine, whereas interference describes the pattern created by constructive and destructive interference. This article studies sound wave interference. Wavefront representation helps see interference. The chapter finishes with a discussion of beats, a fascinating phenomenon resulting from two waves with slightly differing frequencies interacting to create a throbbing pattern.

REVIEW QUESTIONS

1. Explain the difference between a pulse wave and a periodic wave. Provide examples of each.

2. What does "dimensionality" mean in the context of waves, and how does it affect wave behavior?
3. Describe the concept of wave polarization and its significance in wave propagation.
4. How is wave speed determined, and what factors can influence the speed of a wave?
5. Define "simple harmonic motion" and explain its relevance to wave representation.
6. Derive the harmonic wave equation and explain its significance in describing wave behavior mathematically.
7. Discuss multi-dimensional waves with a focus on sound waves. How do sound waves propagate through different media?
8. Differentiate between wavefronts and rays in the context of multi-dimensional waves.
9. What is wave intensity, and how does it relate to the amplitude of a wave?
10. Describe the Doppler Effect and its various scenarios: stationary source and observer, moving source, and moving observer.
11. Explain the concept of superposition and interference. How does interference occur, and what are some specific cases of interference discussed in the chapter?

REFERENCES

1. Ablowitz, M. J., & Zeppetella, A. (1979). Explicit solutions of Fisher's equation for a special wave speed. *Bulletin of Mathematical Biology*, *41*(6), 835–840.
2. Adelman, S. A., & Doll, J. D. (1976). Generalized Langevin equation approach for atom/solid-surface scattering: General formulation for classical scattering off harmonic solids. *The Journal of chemical physics*, *64*(6), 2375–2388.
3. Afandi, Z. (2018, April). Development of gravity acceleration measurement using simple harmonic motion pendulum method based on digital technology and photogate sensor. In *IOP Conference Series: Materials Science and Engineering*. IOP Publishing. (Vol. 335, pp. 012064).
4. Alexander, M. J. (1997). A model of non-stationary gravity waves in the stratosphere and comparison to observations. In *Gravity Wave Processes: Their Parameterization in Global Climate Models*. Berlin, Heidelberg: Springer Berlin Heidelberg. (Vol. 2, pp. 153–168).
5. Ali, A., Seadawy, A. R., & Baleanu, D. (2020). Propagation of harmonic waves in a cylindrical rod via generalized Pochhammer-Chree dynamical wave equation. *Results in Physics*, *17*, 103039.
6. Amrani, D., & Paradis, P. (2010). Use of computer-based data acquisition to teach physics laboratories: case study-simple harmonic motion. *Latin-American Journal of Physics Education*, *4*(3), 6.
7. Baruch, G., Fibich, G., Tsynkov, S., & Turkel, E. (2009). Fourth order schemes for time-harmonic wave equations with discontinuous coefficients. *Commun. Comput. Phys*, *5*, 442–455.
8. Berry, M. V., Chambers, R. G., Large, M. D., Upstill, C., & Walmsley, J. C. (1980). Wavefront dislocations in the Aharonov-Bohm effect and its water wave analogue. *European Journal of Physics*, *1*(3), 154.
9. Besieris, I. M., Shaarawi, A. M., & Ziolkowski, R. W. (1989). A bidirectional traveling plane wave representation of exact solutions of the scalar wave equation. *Journal of Mathematical Physics*, *30*(6), 1254–1269.
10. Bolotovskiĭ, B. M., & Ginzburg, V. L. (1972). The Vavilov-Cerenkov effect and the Doppler effect in the motion of sources with superluminal velocity in vacuum. *Soviet Physics Uspekhi*, *15*(2), 184.
11. Bos, L., & Slawinski, M. A. (2010). Elastodynamic equations:

characteristics, wavefronts and rays. *Quarterly Journal of Mechanics and Applied Mathematics*, *63*(1), 23–38.

12. Brambilla, M. (2002). Quasi-local'wave equations in toroidal geometry with applications to fast wave propagation and absorption at high harmonics of the ion cyclotron frequency. *Plasma Physics and Controlled Fusion*, *44*(11), 2423.
13. Buchenau, U., Galperin, Y. M., Gurevich, V. L., Parshin, D. A., Ramos, M. A., & Schober, H. R. (1992). Interaction of soft modes and sound waves in glasses. *Physical Review B*, *46*(5), 2798.
14. Cally, P. S., & Goossens, M. (2008). Three-dimensional MHD wave propagation and conversion to Alfvén waves near the solar surface. I. Direct numerical solution. *Helioseismology, Asteroseismology, and MHD Connections*, 4, 251–265.
15. Censor, D., & Newhouse, V. L. (1988). Generalized Doppler effect: Coherent and incoherent spectra. *The Journal of the Acoustical Society of America*, *83*(6), 2012–2019.
16. Chang, T., Crew, G. B., Hershkowitz, N., Jasperse, J. R., Retterer, J. M., & Winningham, J. D. (1986). Transverse acceleration of oxygen ions by electromagnetic ion cyclotron resonance with broad band left-hand polarized waves. *Geophysical Research Letters*, *13*(7), 636–639.
17. Chrysohoou, C., Angelis, A., Tsitsinakis, G., Spetsioti, S., Nasis, I., Tsiachris, D., ... & Dimitris, T. (2015). Cardiovascular effects of high-intensity interval aerobic training combined with strength exercise in patients with chronic heart failure. A randomized phase III clinical trial. *International Journal of Cardiology*, *179*, 269–274.
18. Cocco, D., Cutler, G., del Rio, M. S., Rebuffi, L., Shi, X., & Yamauchi, K. (2022). Wavefront preserving X-ray optics for Synchrotron and Free Electron Laser photon beam transport systems. *Physics Reports*, *974*, 1–40.
19. Colson, W. (1981). The nonlinear wave equation for higher harmonics in free-electron lasers. *IEEE Journal of Quantum Electronics*, *17*(8), 1417–1427.
20. Coman, R., & Gajewski, D. (2001, June). 3D Traveltime and migration weight computation using wavefront oriented ray tracing. In *63rd EAGE Conference & Exhibition.* European Association of Geoscientists & Engineers. (Vol. 2, pp. 10–15).
21. Comtois, P., Kneller, J., & Nattel, S. (2005). Of circles and spirals:

bridging the gap between the leading circle and spiral wave concepts of cardiac reentry. *EP Europace*, *7*(s2), S10-S20.

22. Cordesses, L. (2004). Direct digital synthesis: A tool for periodic wave generation (part 1). *IEEE Signal Processing Magazine*, *21*(4), 50–54.
23. Crampin, S. (1984). An introduction to wave propagation in anisotropic media. *Geophysical Journal International*, *76*(1), 17–28.
24. Creze, M., Nordez, A., Soubeyrand, M., Rocher, L., Maître, X., & Bellin, M. F. (2018). Shear wave sonoelastography of skeletal muscle: basic principles, biomechanical concepts, clinical applications, and future perspectives. *Skeletal Radiology*, *47*, 457–471.
25. Daigle, G. A., Piercy, J. E., & Embleton, T. F. W. (1978). Effects of atmospheric turbulence on the interference of sound waves near a hard boundary. *The Journal of the Acoustical Society of America*, *64*(2), 622–630.
26. Davys, J. W., Hosking, R. J., & Sneyd, A. D. (1985). Waves due to a steadily moving source on a floating ice plate. *Journal of Fluid Mechanics*, *158*, 269–287.
27. Dawson, S. V., & Elliott, E. A. (1977). Wave-speed limitation on expiratory flow-a unifying concept. *Journal of Applied Physiology*, *43*(3), 498–515.
28. Deckers, E., Atak, O., Coox, L., D'Amico, R., Devriendt, H., Jonckheere, S., ... & Desmet, W. (2014). The wave based method: An overview of 15 years of research. *Wave Motion*, *51*(4), 550–565.
29. Deng, M. (2003). Analysis of second-harmonic generation of Lamb modes using a modal analysis approach. *Journal of Applied Physics*, *94*(6), 4152–4159.
30. Devaney, A. J., & Wolf, E. (1974). Multipole expansions and plane wave representations of the electromagnetic field. *Journal of Mathematical Physics*, *15*(2), 234–244.
31. Di Benedetto, H., Sauzeat, C., Bilodeau, K., Buannic, M., Mangiafico, S., Nguyen, Q. T., ... & Van Rompu, J. (2011). General overview of the time-temperature superposition principle validity for materials containing bituminous binder. *International Journal of Roads and Airports*, *1*(1), 35–52.
32. Dietz, R. S. (1963). Wave-base, marine profile of equilibrium, and wave-built terraces: a critical appraisal. *Geological Society of America Bulletin*, *74*(8), 971–990.

33. Dimas, A., Suparmi, A., Sarwanto, S., & Nugraha, D. A. (2018, September). Analysis multiple representation skills of high school students on simple harmonic motion. In *AIP Conference Proceedings*. AIP Publishing. (Vol. 201, pp. 987–990).
34. Dwivedi, A. K., Sharma, A., Singh, A. K., & Singh, V. (2021). Metamaterial inspired dielectric resonator MIMO antenna for isolation enhancement and linear to circular polarization of waves. *Measurement*, *182*, 109681.
35. Faber, P., Martínez, S. M., Prieto, M. R. C., Ruiz, J. S., Velasco, J. A. P., Arauz, P. L., ... & Expósito, M. V. (2006). Process-oriented terminology management in the domain of Coastal Engineering. *Terminology. International Journal of Theoretical and Applied Issues in Specialized Communication*, *12*(2), 189–213.
36. Felsen, L. (1984). Progressing and oscillatory waves for hybrid synthesis of source excited propagation and diffraction. *IEEE Transactions on Antennas and Propagation*, *32*(8), 775–796.
37. Feng, J., & Khir, A. W. (2010). Determination of wave speed and wave separation in the arteries using diameter and velocity. *Journal of biomechanics*, *43*(3), 455–462.
38. Fernández, I. S., Chapman, K. E., Peters, J. M., Kothare, S. V., Nordli Jr, D. R., Jensen, F. E., ... & Loddenkemper, T. (2013). The tower of Babel: survey on concepts and terminology in electrical status epilepticus in sleep and continuous spikes and waves during sleep in North America. *Epilepsia*, *54*(4), 741–750.
39. Fliss, S., & Joly, P. (2016). Solutions of the time-harmonic wave equation in periodic waveguides: asymptotic behaviour and radiation condition. *Archive for Rational Mechanics and Analysis*, *219*, 349–386.
40. Fujimoto, S., Mizuno, R., Saito, Y., & Nakamura, S. (2004). Clinical application of wave intensity for the treatment of essential hypertension. *Heart and vessels*, *19*, 19–22.
41. Galeriu, C., Edwards, S., & Esper, G. (2014). An Arduino investigation of simple harmonic motion. *The Physics Teacher*, *52*(3), 157–159.
42. Galindo-Rodríguez, J. I., & Silva-Ortigoza, G. (2021). Wavefronts, light rays and caustic associated with the refraction of a plane wavefront by a conospherical lens. *Journal of Optics*, *23*(12), 125605.
43. Garfinkle, D. (2002). Harmonic coordinate method for simulating

generic singularities. *Physical Review D*, *65*(4), 044029.

44. Gouesbet, G., Lock, J. A., & Gréhan, G. (1995). Partial-wave representations of laser beams for use in light-scattering calculations. *Applied Optics*, *34*(12), 2133–2143.
45. Gregory, J. M., Griffies, S. M., Hughes, C. W., Lowe, J. A., Church, J. A., Fukimori, I., ... & van de Wal, R. S. (2019). Concepts and terminology for sea level: Mean, variability and change, both local and global. *Surveys in Geophysics*, *40*, 1251–1289.
46. Grilli, S. T., Guyenne, P., & Dias, F. (2001). A fully non-linear model for three-dimensional overturning waves over an arbitrary bottom. *International Journal for Numerical Methods in Fluids*, *35*(7), 829–867.
47. Guo, W., & Guo, B. Z. (2013). Parameter estimation and non-collocated adaptive stabilization for a wave equation subject to general boundary harmonic disturbance. *IEEE Transactions on Automatic Control*, *58*(7), 1631–1643.
48. Habib, A., Shelke, A., Pluta, M., Pietsch, U., Kundu, T., & Grill, W. (2012, May). Scattering and attenuation of surface acoustic waves and surface skimming longitudinal polarized bulk waves imaged by Coulomb coupling. In *AIP Conference Proceedings*. American Institute of Physics. (Vol. 1433, pp. 247–250).
49. Hasselmann, K., Munk, W., & MacDonald, G. J. (1963). Bispectra of ocean waves. In *Symposium on time series analysis*. New York. (Vol. 2, pp. 125–139).
50. Hirasawa, T., & Stauder, W. (1965). On the seismic body waves from a finite moving source. *Bulletin of the Seismological Society of America*, *55*(2), 237–262.
51. Hirata, K., Kawakami, M., & O'Rourke, M. F. (2006). Pulse wave analysis and pulse wave velocity a review of blood pressure interpretation 100 years after Korotkov. *Circulation Journal*, *70*(10), 1231–1239.
52. Hirsch, E., Caraballo, R., Bernardina, B. D., Loddenkemper, T., & Zuberi, S. M. (2019). Encephalopathy related to Status Epilepticus during slow Sleep: from concepts to terminology. *Epileptic Disorders*, *21*, S5-S12.
53. Hosokawa, K., Miyoshi, Y., Ozaki, M., Oyama, S. I., Ogawa, Y., Kurita, S., ... & Fujii, R. (2020). Multiple time-scale beats in aurora: Precise

orchestration via magnetospheric chorus waves. *Scientific Reports*, *10*(1), 3380.

54. Hudson, J. A. (1981). Wave speeds and attenuation of elastic waves in material containing cracks. *Geophysical Journal International*, *64*(1), 133–150.
55. Juárez-Reyes, S. A., Ortega-Vidals, P., & Silva-Ortigoza, G. (2015). Wavefronts, light rays and caustic associated with the refraction of a spherical wave by two interfaces: the axicon and the plano-convex parabolic lenses. *Journal of Optics*, *17*(6), 065604.
56. Kaschwich, T., & Gajewski, D. (2003, June). Wavefront-oriented ray tracing in 3D anisotropic media. In *65th EAGE Conference & Exhibition.* European Association of Geoscientists & Engineers. (Vol. 2, pp. 5–6).
57. Kawase, H. (1996). The cause of the damage belt in Kobe:"The basin-edge effect," constructive interference of the direct S-wave with the basin-induced diffracted/Rayleigh waves. *Seismological Research Letters*, *67*(5), 25–34.
58. Keyrouz, S., & Caratelli, D. (2016). Dielectric resonator antennas: basic concepts, design guidelines, and recent developments at millimeter-wave frequencies. *International Journal of Antennas and Propagation*, *20, 18–25.*
59. Khalid, M., Hadi, F., & Rahman, A. U. (2021). Modulation of multi-dimensional waves in anisotropic magnetized plasma. *The European Physical Journal Plus*, *136*, 1–11.
60. Khir, A. W., O'brien, A., Gibbs, J. S. R., & Parker, K. H. (2001). Determination of wave speed and wave separation in the arteries. *Journal of Biomechanics*, *34*(9), 1145–1155.
61. Kim, U. H., Mizumoto, T., Ogata, T., & Okuno, H. G. (2011, September). Improvement of speaker localization by considering multipath interference of sound wave for binaural robot audition. In *2011 IEEE/RSJ International Conference on Intelligent Robots and Systems*. IEEE. (Vol. 2, pp. 2910–2915).
62. Kinchin, J. (2016). Using Tracker to prove the simple harmonic motion equation. *Physics Education*, *51*(5), 053003.
63. Kingsley, J. D., Tai, Y. L., Vaughan, J. A., & Mayo, X. (2017). High-intensity interval cycling exercise on wave reflection and pulse wave velocity. *The Journal of Strength & Conditioning Research*, *31*(5), 1313–1320.

64. Kinnunen, J. V., Piitulainen, H., & Piirainen, J. M. (2019). Neuromuscular adaptations to short-term high-intensity interval training in female ice-hockey players. *The Journal of Strength & Conditioning Research*, *33*(2), 479–485.
65. Kiselev, V. A., Razbirin, B. S., & Uraltsev, I. N. (1975). Additional waves and Fabry-Perot interference of photoexcitons (polaritons) in thin II–VI crystals. *Physica Status Solidi (B)*, *72*(1), 161–172.
66. Knobel, R. (2000). *An Introduction to the Mathematical Theory of Waves*. American Mathematical Soc. (Vol. 3, pp. 12–15).
67. Kolokolov, A. A., & Skrotskiĭ, G. V. (1992). Interference of reactive components of an electromagnetic field. *Soviet Physics Uspekhi*, *35*(12), 1089.
68. Kovachy, T., Asenbaum, P., Overstreet, C., Donnelly, C. A., Dickerson, S. M., Sugarbaker, A., ... & Kasevich, M. A. (2015). Quantum superposition at the half-metre scale. *Nature*, *528*(7583), 530–533.
69. Kress, R. (1979). On the limiting behaviour of solutions to boundary integral equations associated with time harmonic wave equations for small frequencies. *Mathematical Methods in the Applied Sciences*, *1*(1), 89–100.
70. Kurner, T., Cichon, D. J., & Wiesbeck, W. (1993). Concepts and results for 3D digital terrain-based wave propagation models: An overview. *IEEE Journal on Selected Areas in Communications*, *11*(7), 1002–1012.
71. Kuvshinov, B. N., & Mulder, W. A. (2006). The exact solution of the time-harmonic wave equation for a linear velocity profile. *Geophysical Journal International*, *167*(2), 659–662.
72. Lan, K. C., & Shih, W. Y. (2012, April). Using simple harmonic motion to estimate walking distance for waist-mounted PDR. In *2012 IEEE Wireless Communications and Networking Conference (WCNC)*. IEEE. (Vol. 2, pp. 2445–2450).
73. Langley, R. S., & Bercin, A. N. (1994). Wave intensity analysis of high frequency vibrations. *Philosophical Transactions of the Royal Society of London. Series A: Physical and Engineering Sciences*, *346*(1681), 489–499.
74. Lin, J., Chen, C. S., Liu, C. S., & Lu, J. (2016). Fast simulation of multi-dimensional wave problems by the sparse scheme of the method of fundamental solutions. *Computers & Mathematics with Applications*, *72*(3), 555–567.

75. Liu, S., Fu, Z., Liu, S., & Zhao, Q. (2001). Jacobi elliptic function expansion method and periodic wave solutions of nonlinear wave equations. *Physics Letters A*, *289*(1–2), 69–74.
76. Luo, Y., Chen, Y. J., Zhu, Y. Z., Li, W. Y., & Zhang, Q. (2020). Doppler effect and micro-Doppler effect of vortex-electromagnetic-wave-based radar. *IET Radar, Sonar & Navigation*, *14*(1), 2–9.
77. Luostari, T., & Huttunen, T. (2009). time harmonic wave equation. *Highly Oscillatory Problems*, (366), 127.
78. Ma, W. X., Zhou, R., & Gao, L. (2009). Exact one-periodic and two-periodic wave solutions to Hirota bilinear equations in (2+ 1) dimensions. *Modern Physics Letters A*, *24*(21), 1677–1688.
79. Martirosyan, N. A., Dolukhanyan, S. K., & Marzhanov, A. G. (1984). Experimental observation of the nonuniqueness of stationary combustion in systems with parallel reactions. *Combust., Explos. Shock Waves (Engl. Transl.); (The United States)*, *19*(6), 154–160.
80. McCumber, D. E. (1968). Effect of ac impedance on dc voltage-current characteristics of superconductor weak-link junctions. *Journal of Applied Physics*, *39*(7), 3113–3118.
81. Miles, J. W. (1971). Internal waves generated by a horizontally moving source. *Geophysical Fluid Dynamics*, *2*(1), 63–87.
82. Misquitta, C. M., Mack, D. P., & Grover, A. K. (1999). Sarco/endoplasmic reticulum Ca2+ (SERCA)-pumps: link to heart beats and calcium waves. *Cell calcium*, *25*(4), 277–290.
83. Mogensen, P., Pajukoski, K., Tiirola, E., Vihriala, J., Lahetkangas, E., Berardinelli, G., ... & Cattoni, A. F. (2014, May). Centimeter-wave concept for 5G ultra-dense small cells. In *2014 IEEE 79th Vehicular Technology Conference (VTC Spring)* IEEE. (Vol. 2, pp. 1–6).
84. Nakamura, A. (1979). A direct method of calculating periodic wave solutions to nonlinear evolution equations. I. Exact two-periodic wave solution. *Journal of the Physical Society of Japan*, *47*(5), 1701–1705.
85. Nakamura, A. (1980). A direct method of calculating periodic wave solutions to nonlinear evolution equations. II. Exact one-and two-periodic wave solution of the coupled bilinear equations. *Journal of the Physical Society of Japan*, *48*(4), 1365–1370.
86. Nédélec, J. C. (2001). *Acoustic and Electromagnetic Equations: Integral Representations for Harmonic Problems.* New York: Springer. (Vol. 144, pp. 58–60).

87. Nezlin, M. V. (1976). Negative-energy waves and the anomalous Doppler effect. *Uspekhi Fizicheskikh Nauk, 120*, 481–495.

88. Niki, K., Sugawara, M., Chang, D., Harada, A., Okada, T., Sakai, R., ... & Mumford, C. E. (2002). A new noninvasive measurement system for wave intensity: evaluation of carotid arterial wave intensity and reproducibility. *Heart and Vessels*, *17*(1), 12–21.

89. Noel, M. W., & Stroud Jr, C. R. (1996). Excitation of an atomic electron to a coherent superposition of macroscopically distinct states. *Physical Review Letters*, *77*(10), 1913.

90. Pambayun, B., Wirjawan, J. V., Wijaya, A., Untung, G. B., & Pratidhina, E. (2019). Designing mobile learning app to help high school students to learn simple harmonic motion. *International Journal on Social and Education Sciences*, *1*(1), 24–29.

91. Parker, K. H. (2009). An introduction to wave intensity analysis. *Medical & Biological Engineering & Computing*, *47*, 175–188.

92. Pavić, G. (1976). Measurement of structure borne wave intensity, Part I: Formulation of the methods. *Journal of Sound and Vibration*, *49*(2), 221–230.

93. Pedlosky, J. (2003). *Waves in the Ocean and Atmosphere: Introduction to Wave Dynamics.* Berlin: Springer. (Vol. 260, pp. 25–30).

94. Polifke, W., Paschereit, C. O., & Döbbeling, K. (2001). Constructive and destructive interference of acoustic and entropy waves in a premixed combustor with a choked exit. *Int. J. Acoust. Vib*, *6*(3), 135–146.

95. Pu, X., Meng, Q., & Shi, Z. (2020). Experimental studies on surface-wave isolation by periodic wave barriers. *Soil Dynamics and Earthquake Engineering*, *130*, 106000.

96. Qiu, S., Liu, T., Ding, Y., Liu, Z., Chen, L., & Ren, Y. (2022). Rotational Doppler effect with vortex beams: Fundamental mechanism and technical progress. *Frontiers in Physics*, *10*, 938593.

97. Raman, C. V., & Nagendra Nath, N. S. (1936, January). The diffraction of light by high frequency sound waves: Part III: Doppler effect and coherence phenomena. In *Proceedings of the Indian Academy of Sciences-Section A.* New Delhi: Springer India. (Vol. 3, pp. 75–84).

98. Ramsey, M. W., & Sugawara, M. (1997). Arterial wave intensity and ventriculoarterial interaction. *HEART AND VESsels*, 10(4), 128–134.

99. Rawlinson, N., Hauser, J., & Sambridge, M. (2008). Seismic ray tracing and wavefront tracking in laterally heterogeneous media. *Advances in*

Geophysics, *49*, 203–273.

100. Rayces, J. L. (1964). Exact relation between wave aberration and ray aberration. *Optica Acta: International Journal of Optics*, *11*(2), 85–88.
101. Raymond, W. H., & Kuo, H. L. (1984). A radiation boundary condition for multi-dimensional flows. *Quarterly Journal of the Royal Meteorological Society*, *110*(464), 535–551.
102. Reed, E. J., Soljačić, M., & Joannopoulos, J. D. (2003). Reversed Doppler effect in photonic crystals. *Physical Review Letters*, *91*(13), 133901.
103. Riyanti, C. D., Erlangga, Y. A., Plessix, R. E., Mulder, W. A., Vuik, C., & Oosterlee, C. (2006). A new iterative solver for the time-harmonic wave equation. *Geophysics*, *71*(5), E57–E63.
104. Rosakis, A. J., Samudrala, O., & Coker, D. (1999). Cracks faster than the shear wave speed. *Science*, *284*(5418), 1337–1340.
105. Roshchupkin, S. P., & Lysenko, O. B. (1999). Spontaneous interference bremsstrahlung effect in the scattering of a relativistic electron by a nucleus in the field of two light waves. *Journal of Experimental and Theoretical Physics*, *89*, 647–663.
106. Rousseaux, G., Mathis, C., Maïssa, P., Philbin, T. G., & Leonhardt, U. (2008). Observation of negative-frequency waves in a water tank: a classical analogue to the Hawking effect?. *New Journal of Physics*, *10*(5), 053015.
107. Schulkes, R. M. S. M., Hosking, R. J., & Sneyd, A. D. (1987). Waves due to a steadily moving source on a floating ice plate. Part 2. *Journal of Fluid Mechanics*, *180*, 297–318.
108. Shao, Y. L., & Faltinsen, O. M. (2014). A harmonic polynomial cell (HPC) method for 3D Laplace equation with application in marine hydrodynamics. *Journal of Computational Physics*, *274*, 312–332.
109. Shen, Y., Tian, B., Zhang, C. R., Tian, H. Y., & Liu, S. H. (2021). Breather-wave, periodic-wave and traveling-wave solutions for a (2+1)-dimensional extended Boiti–Leon–Manna–Pempinelli equation for an incompressible fluid. *Modern Physics Letters B*, *35*(15), 2150261.
110. Somroob, S., & Wattanakasiwich, P. (2017, September). Investigating student understanding of simple harmonic motion. In *Journal of Physics: Conference Series*. IOP Publishing. (Vol. 901, pp. 012123).
111. Spengler, J. F., Jekel, M., Christensen, K. T., Adrian, R. J., Hawkes, J. J., & Coakley, W. T. (2000). Observation of yeast cell movement

and aggregation in a small-scale MHz-ultrasonic standing wave field. *Bioseparation*, *9*, 329–341.

112. Sugawara, M., Niki, K., Ohte, N., Okada, T., & Harada, A. (2009). Clinical usefulness of wave intensity analysis. *Medical & Biological Engineering & Computing*, *47*, 197–206.
113. Sun, Y. H., Anderson, T. J., Parker, K. H., & Tyberg, J. V. (2000). Wave-intensity analysis: a new approach to coronary hemodynamics. *Journal of Applied Physiology*, *89*(4), 1636–1644.
114. Suppes, P., & Han, B. (2000). Brain-wave representation of words by superposition of a few sine waves. *Proceedings of the National Academy of Sciences*, *97*(15), 8738–8743.
115. Svirkunov, P. N., & Kalashnik, M. V. (2014). Phase patterns of dispersive waves from moving localized sources. *Physics-Uspekhi*, *57*(1), 80.
116. Sweeney, M. O., Hellkamp, A. S., van Bommel, R. J., Schalij, M. J., Borleffs, C. J. W., & Bax, J. J. (2014). QRS fusion complex analysis using wave interference to predict reverse remodeling during cardiac resynchronization therapy. *Heart Rhythm*, *11*(5), 806–813.
117. Thevenot, M. M., & Kraus, N. C. (1995). Longshore sand waves at Southampton Beach, New York: observation and numerical simulation of their movement. *Marine Geology*, *126*(1–4), 249–269.
118. Thoads, T. R., & Roedel, R. J. (1999, November). The wave concept inventory-a cognitive instrument based on Bloom's taxonomy. In *FIE'99 Frontiers in Education. 29th Annual Frontiers in Education Conference. Designing the Future of Science and Engineering Education. Conference Proceedings (IEEE Cat. No. 99CH37011.* IEEE. (Vol. 3, pp. 13–14).
119. Tijdeman, H. (1975). On the propagation of sound waves in cylindrical tubes. *Journal of Sound and Vibration*, *39*(1), 1–33.
120. Tomonaga, S. I. (1950). Remarks on Bloch's method of sound waves applied to many-fermion problems. *Progress of Theoretical Physics*, *5*(4), 544–569.
121. Triana, C. A., & Fajardo, F. (2011). The influence of spring length on the physical parameters of simple
122. Tsurutani, B. T., & Lakhina, G. S. (1997). Some basic concepts of wave-particle interactions in collisionless plasmas. *Reviews of Geophysics*, *35*(4), 491–501.

123. Tucker, M. J. (1950). Surf beats: Sea waves of 1 to 5 min. period. *Proceedings of the Royal Society of London. Series A. Mathematical and Physical Sciences*, *202*(1071), 565–573.
124. Tyberg, J. V., Davies, J. E., Wang, Z., Whitelaw, W. A., Flewitt, J. A., Shrive, N. G., ... & Wang, J. J. (2009). Wave intensity analysis and the development of the reservoir–wave approach. *Medical & Biological Engineering & Computing*, *47*, 221–232.
125. Überall, H. (1998). Interference and steady-state scattering of sound waves. *Handbook of Acoustics (MJ Crocker (Ed.); Chap. 4*, 47–60.
126. Van Kruining, K. C., Cameron, R. P., & Götte, J. B. (2018). Superpositions of up to six plane waves without electric-field interference. *Optica*, *5*(9), 1091–1098.
127. Vinh, P. C., & Ogden, R. W. (2004). On formulas for the Rayleigh wave speed. *Wave Motion*, *39*(3), 191–197.
128. Vinje, V., Iversen, E., Åstebøl, K., & Gjøystdal, H. (1996). Estimation of multivalued arrivals in 3d models using wavefront construction—Part I. *Geophysical Prospecting*, *44*(5), 819–842.
129. Voti, R. L., & Bertolotti, M. (2021). Thermal waves emitted by moving sources and the Doppler effect. *International Journal of Heat and Mass Transfer*, *176*, 121098.
130. Wang, M., & Li, X. (2005). Applications of F-expansion to periodic wave solutions for a new Hamiltonian amplitude equation. *Chaos, Solitons & Fractals*, *24*(5), 1257–1268.
131. Waterhouse, R. V. (1955). Interference patterns in reverberant sound fields. *The Journal of the Acoustical Society of America*, *27*(2), 247–258.
132. Wei, J., Qiu, J., Li, T., Huang, Y., Qiao, Z., Cao, J., ... & Wang, G. (2021). Cloud and precipitation interference by strong low-frequency sound wave. *Science China Technological Sciences*, *64*(2), 261–272.
133. Wittmann, M., Steinberg, R. N., & Redish, E. F. (2003). Understanding and affecting student reasoning about sound waves. *International Journal of Science Education*, *25*(8), 991–1013.
134. Wu, R. S., & Aki, K. (1988). Introduction: Seismic wave scattering in three-dimensionally heterogeneous earth. In *Scattering and Attenuations of Seismic Waves, Part.* Basel: Birkhäuser Basel. (Vol. 2, pp. 1–6).
135. Xu, C., Chen, Z. G., Zhang, G., Ma, G., & Wu, Y. (2021). Multi-

dimensional wave steering with higher-order topological phononic crystal. *Science Bulletin*, *66*(17), 1740–1745.

136. Yan, Z., & Zhang, H. (2001). New explicit solitary wave solutions and periodic wave solutions for Whitham–Broer–Kaup equation in shallow water. *Physics Letters A*, *285*(5–6), 355–362.
137. Yeung, R. W., & Nguyen, T. C. (1999). Waves generated by a moving source in a two-layer ocean of finite depth. *Journal of Engineering Mathematics*, *35*, 85–107.
138. Zheng, D., Lyu, Y. L., & Wu, K. (2019). Longitudinally slotted SIW leaky-wave antenna for low cross-polarization millimeter-wave applications. *IEEE Transactions on Antennas and Propagation*, *68*(2), 656–664.
139. Zhou, Y., Wang, M., & Wang, Y. (2003). Periodic wave solutions to a coupled KdV equations with variable coefficients. *Physics Letters A*, *308*(1), 31–36.
140. Ziemkiewicz, D., & Zielińska-Raczyńska, S. (2015). Cherenkov radiation combined with the complex Doppler effect in left-handed metamaterials. *JOSA B*, *32*(8), 1637–1644.
141. Zolotukhina, N. A. (1975). On the excitation of Alfven waves in the magnetosphere by a moving source. *On the Excitation of Alfven Waves in the Magnetosphere by a Moving Source Transl. Into ENGLISH from Issled. po Geomagn. Aeron. Fiz. Solntsa (Irkutsk*, *34*, 20–23.

CHAPTER 2

Wave Phenomena in Nature

Contents

2.1. INTRODUCTION

Waves are one of the most fundamental and pervasive phenomena that occur in the natural world. They play an important part in the formation of our environment and the governance of various physical processes. In this chapter, we dig into the fascinating realm of waves, investigating their definition, their properties, and their role in the natural world. Let's get started on our voyage of discovery to learn about how waves have an effect on the world around us. In order to fully understand the significance of waves in the natural world, we must first get a firm grasp on their literal meaning. Waves are characterized by disturbances or oscillations that move through a medium or space and carry energy without moving matter. Waves can be thought of as moving disturbances or oscillations. Because of this one-of-a-kind quality, waves can travel great distances while simultaneously carrying energy and data from one place to another (Jess et al., 2012).

Wave phenomena is an interesting branch of physics that investigates the characteristics and behaviors of waves as they travel across different mediums. Scientists have deciphered the complex dynamics of wave motion in their endeavor to understand the fundamental principles that control these fascinating natural events. The study of wave behavior has significant repercussions in various scientific disciplines, including acoustics, optics, seismology, meteorology, and astrophysics (Fratamico et al., 2016).

The awe-inspiring nature of the natural world may be summed up in one word: waves. Waves can be found all around us, in a wide variety of guises, from the tiniest of ripples on a calm lake to the enormous seismic waves that alter the terrain of our planet. Their relevance extends far and wide, affecting many different facets of life. Sound waves are essential to effective communication in the atmosphere among all life forms. Birds can communicate with one another through beautiful songs, whales can communicate across enormous distances using complex cries, and insects emit noises to attract mates or mark their territory. Sound waves are also essential to humans since they enable us to communicate verbally, create music, and share knowledge (McElroy et al., 2003).

The fascination with the ocean waves is indisputable because of the fascinating patterns they create when they smash upon the shore. In addition to their aesthetic appeal, they have a role in forming weather patterns and regulating the Earth's climate. Even though they are relatively uncommon, tsunamis are a striking reminder of the enormous force trapped within sea waves, power that can potentially have catastrophic effects (Zaikin &

Zhabotinsky, 1970). Light waves, a form of electromagnetic wave, have a very important function in the overall structure and operation of the cosmos. Without them, our world would be shrouded in darkness, and we would be unable to appreciate the natural splendor surrounding us. Light is required for photosynthesis in plants, which results in the production of energy that is necessary for continued life. Light waves convey information from faraway stars and provide astronomers with this data, which is essential for their understanding of the universe. Light waves travel beyond Earth (Dörfler et al., 2023).

2.2. TYPES OF WAVES IN NATURE

The classification of waves in the natural world encompasses a wide range of phenomena that exert an influence on our environment and contribute to the formation and transformation of the surrounding world. Mechanical waves, such as water waves and seismic waves, necessitate a medium through which they can propagate, thereby giving rise to the tranquil undulations of ocean waves and the formidable impact of earthquakes. In contrast, electromagnetic waves propagate through the absence of matter in space and encompass various types, such as light waves, radio waves, and others (Tsai et al., 2016). Electromagnetic waves, which propagate through space, carry energy in the form of discrete packets known as photons. These waves play a crucial role in facilitating human visual perception, communication, and comprehension of the cosmos. In conclusion, matter waves serve as a means to elucidate the distinct quantum characteristics exhibited by particles, thereby showcasing a remarkable duality between particles and waves. These various wave classifications enhance our comprehension of the natural environment, revealing its intricate nature and interrelatedness (Gruszecki et al., 2021).

2.2.1. Mechanical Waves

Mechanical waves are a captivating classification of waves that necessitate a medium in order to propagate. In contrast to electromagnetic waves, which can propagate through a vacuum, mechanical waves necessitate the involvement of particles within a medium to facilitate the transmission of energy. Let's explore the characteristics of mechanical waves and some captivating examples found in nature. Tongchai et al., 2011).

2.2.1.1. Characteristics of Mechanical Waves

Mechanical waves, fundamentally, arise from perturbations within a medium, leading to the oscillatory motion of particles around their respective equilibrium positions (Figure 2.1). These disturbances facilitate energy transfer between different regions without any overall displacement of matter. The medium functions as a channel for energy transmission, with particles reciprocally undergoing oscillations, facilitating energy transfer as the wave propagates through the material (Sumi & Ashida, 2003).

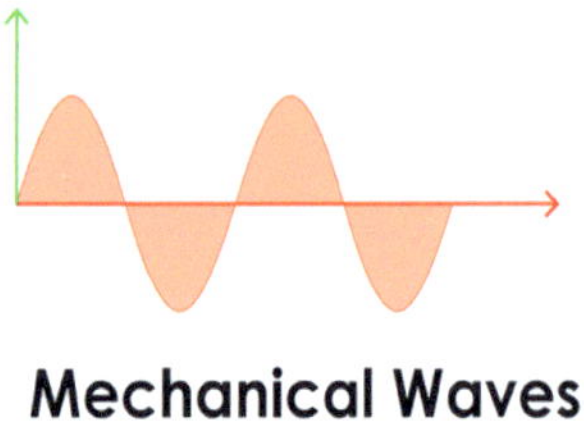

Figure 2.1. Representation of a mechanical wave (Young Wonks, Creative Commons License).

One of the primary distinguishing features of mechanical waves is their capacity to be categorized into two principal classifications: transverse waves and longitudinal waves.

- Transverse waves are characterized by the perpendicular oscillation of particles in the medium concerning the direction of wave propagation. Consider a hypothetical scenario in which a wave propagates along a rope, causing the rope particles to undergo vertical oscillations as it moves upward and downward. Water waves are a well-known illustration of a transverse mechanical wave, characterized by the presence of crests and troughs on the surface of the water as the wave propagates (Wang et al., 2020).
- Longitudinal waves involve the oscillation of particles in the medium in a parallel manner to the direction of wave propagation. The translational motion of particles is aligned with the wave's propagation direction. This phenomenon becomes evident when a spring undergoes compression and expansion, resulting in the particles of the spring oscillating in a reciprocating manner aligned with the direction of the wave. One notable illustration of a longitudinal mechanical wave can be observed in the seismic

wave produced by an earthquake, wherein the particles of the Earth undergo oscillations in alignment with the direction of wave propagation as it traverses through the ground (Scarpa et al., 2011).

2.2.1.2. Examples of Mechanical Waves in Nature

The subsequent instances illustrate the presence of mechanical waves in the natural environment:

- Water Waves: One of the most captivating illustrations of mechanical waves in the natural world is the phenomenon of waves observed on the surface of bodies of water. Transverse waves, whether they manifest as gentle ripples in a serene pond or as powerful crashing waves in the vast expanse of an ocean, arise from the intricate interplay between wind and water. The transfer of energy from the wind to the surface of the water gives rise to the formation of visually appealing undulations, which propagate throughout the water's expanse (Iversen et al., 2018).
- Seismic waves are powerful vibrations propagating through the Earth's interior (Figure 2.2). These vibrations are a direct consequence of earthquakes. The waves in question are predominantly longitudinal, specifically called Primary (P) and Secondary (S) waves. Due to their higher velocity and ability to propagate through solid and liquid mediums, P-waves exhibit the earliest arrival time in seismic events. S-waves, which have a slower velocity and are restricted to propagating solely through solid materials, follow the arrival of P-waves. The examination of seismic waves aids seismologists in comprehending the internal composition of the Earth and monitoring seismic events (Malladi et al., 2015).

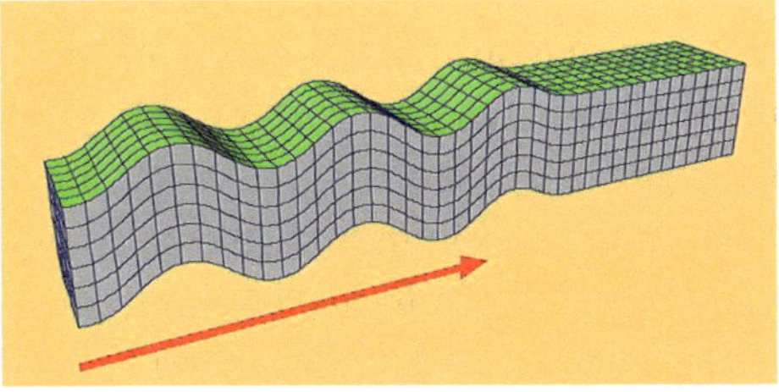

Figure 2.2. Representation of the seismic wave (Source: Britannica, Creative Commons License).

These examples only provide a superficial glimpse into the extensive array of mechanical waves in the natural environment. Mechanical waves are omnipresent in our environment, manifesting in various forms such as captivating sound wave melodies and intricate wave interactions within Earth's atmosphere. Their existence and importance permeate every facet of our surroundings (Tongchai et al., 2009).

2.2.2. Electromagnetic Waves

Electromagnetic waves exhibit a remarkable and wide variety of characteristics, spanning a broad spectrum of energies. In contrast to mechanical waves, electromagnetic waves possess the ability to propagate without the need for a medium, enabling them to traverse the vacuum of space. These waves are generated through the oscillation of magnetic and electric fields and serve a fundamental purpose in the operation of the cosmos. Let's explore the nature, properties, and captivating examples of electromagnetic waves in nature (Podolefsky & Finkelstein, 2007).

2.2.2.1. Nature of Electromagnetic Waves

The fundamental nature of electromagnetic waves is rooted in the intricate interaction between electric and magnetic fields. The generation of these waves occurs as a result of the acceleration of charged particles, such as electrons. The oscillating charges give rise to a perturbation in the electric field, which subsequently generates a magnetic field through induction. The dynamic alteration of the magnetic field subsequently gives rise to an electric field, and this cyclical phenomenon recurs, leading to the emergence of a self-perpetuating wave that propagates through the spatial medium (Zhao et al., 2020).

One notable characteristic of electromagnetic waves is their capacity to propagate through a vacuum, such as the expanse of outer space. This particular attribute serves as a distinguishing factor between electromagnetic waves and mechanical waves, as the former can propagate without the need for a medium (Figure 2.3). At the same time, the latter necessitates a medium for transmission. Electromagnetic waves can propagate through the immense expanse of space at the velocity of light, an astonishing 299,792 kilometers per second (186,282 miles per second) (Ambroselli et al., 2011).

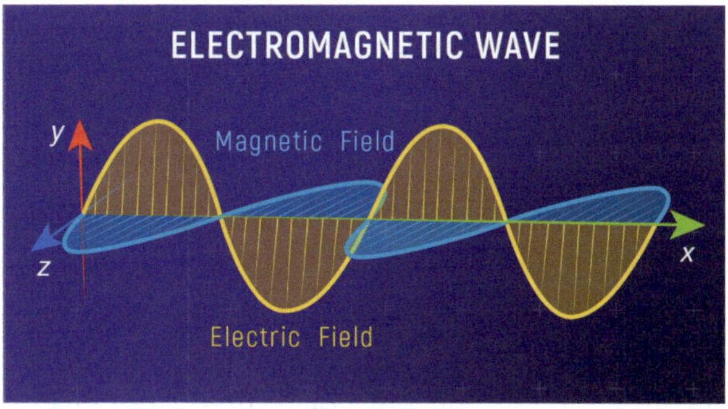

Figure 2.3. Representation of an electromagnetic wave (Source: NASA, Creative Commons License).

The electromagnetic spectrum is a classification system that categorizes electromagnetic waves based on their wavelength and frequency, thereby constituting a distinctive characteristic of these waves. The electromagnetic spectrum encompasses a wide continuum, extending from gamma rays and X-rays with high energy and short wavelengths to visible light and radio waves and finally to microwaves and radio waves with long and low energy (Liang et al., 2016).

2.2.2.2. Examples of Electromagnetic Waves in Nature

The subsequent instances illustrate the presence of mechanical waves in the natural world:

- Light waves are a prominent and captivating illustration of electromagnetic waves in the natural world. The specific range of wavelengths within the electromagnetic spectrum is accountable for facilitating the visual perception of individuals and uncovering the diverse array of colors present in the surrounding environment. The perception of colors is contingent upon the varying wavelengths of light. The wavelength of red light is longer, whereas the wavelength of blue light is shorter. The glorious array of colors exhibited by a rainbow is a magnificent demonstration of the phenomenon of light wave refraction and dispersion as it traverses through atmospheric water droplets (Taflove & Umashankar, 1989).
- Radio waves are classified as electromagnetic radiation characterized by extended wavelengths and reduced frequencies. The utilization of these waves encompasses a wide range of

practical applications, such as wireless communication, television and radio broadcasting, and satellite transmissions. Radio waves have facilitated worldwide communication, fostering connectivity among individuals and facilitating the dissemination of information across extensive geographical areas (Deryck, 1978).

- Microwaves, characterized by wavelengths slightly shorter than those of radio waves, find utility in a range of practical applications, including but not limited to microwave ovens and radar technology. Microwave ovens operate through the generation of electromagnetic waves, which induce the excitation of water molecules, resulting in vibrational motion and subsequent heat production. Microwaves are employed in radar systems for object detection and tracking, encompassing weather phenomena and aerial vehicles (Luo et al., 2022).

The examples above demonstrate electromagnetic waves' vast array and profound importance within natural phenomena. In addition to the visible spectrum, electromagnetic waves span a wide range of energies, encompassing high-energy gamma rays and X-rays utilized in medical imaging, astronomical observations, and radio waves that facilitate global communication (Dong et al., 2019).

2.2.3. Matter Waves

The concept of matter waves, which is highly intriguing within the realm of quantum mechanics, demonstrates the dualistic nature of particles, exhibiting characteristics of both particles and waves (Figure 2.4). The concept of this fascinating duality was initially introduced by Louis de Broglie, who postulated that particles, including electrons and other subatomic particles, demonstrate characteristics akin to waves. This discussion aims to delve into the introductory concepts surrounding matter waves and the intricate quantum phenomena that dictate the behavior of particles. Historically, waves have been conventionally linked to energy characteristics rather than particles. Nevertheless, during the initial decades of the 20th century, physicists explored the enigmatic domain of subatomic particles, where they confronted perplexing occurrences that contradict established principles of classical physics. In his doctoral thesis in 1924, Louis de Broglie postulated that if light could manifest both wave-like and particle-like characteristics, as evidenced by the dual nature of light observed in the renowned double-slit experiment, it is plausible that particles such as electrons could also display analogous duality (Billy et al., 2008).

Figure 2.4. Illustration of a matter wave (Source: Nagwa, Creative Commons License).

De Broglie's hypothesis postulates that a particle exhibiting momentum is accompanied by a corresponding wavelength, which is currently referred to as the de Broglie wavelength (λ). The relationship between the de Broglie wavelength and the momentum of a particle is one of inverse proportionality. This notable revelation introduced the concept of matter waves, wherein particles exhibit wave-like properties alongside their particle-like characteristics (Greiner et al., 2002).

The notion of matter waves is intricately linked to the principles of quantum mechanics, a field of physics that elucidates the characteristics and dynamics of particles at the atomic and subatomic levels. Within the realm of quantum physics, particles deviate from classical trajectories and instead exhibit behavior that can be described using probability distributions. Several significant phenomena serve as examples of the wave-like characteristics of particles and emphasize the significance of matter waves in the field of quantum mechanics (Ginsberg et al., 2007).

- The concept of wave-particle duality pertains to the demonstration of matter waves, which emphasizes the simultaneous existence of particle and wave characteristics. The phenomenon of duality implies that particles can demonstrate characteristics of localized particles and wave-like interference patterns.
- The Heisenberg Uncertainty Principle, a fundamental principle in the field of quantum mechanics, posits that there exists a limitation in accurately determining certain pairs of physical properties, such as position and momentum, concurrently. The wave-like characteristics exhibited by particles heavily influence the incorporation of the principle.

- Electron diffraction experiments demonstrate the manifestation of wave-like properties in electrons. These experiments involve the passage of electrons through a double-slit, resulting in the production of an interference pattern that bears a resemblance to the patterns observed in light wave diffraction. The observed occurrence presents persuasive substantiation supporting the presence of matter waves (Schumm et al., 2005).
- Quantum tunneling is a phenomenon in which matter waves are crucial, enabling particles to traverse energy barriers considered impenetrable in classical physics. The quantum mechanical phenomenon discussed herein has substantial ramifications across diverse disciplines, including electronics and nuclear physics.

The introduction of matter waves has significantly transformed our comprehension of the quantum realm, yielding profound revelations regarding particles' fundamental properties and dynamics. Furthermore, it has established the fundamental principles for contemporary technologies, such as electron microscopes and particle accelerators, which exploit the wave-like characteristics of particles (Müller et al., 2010).

2.3. WATER WAVES

Water waves are intriguing and ever-changing phenomena that manifest on the surfaces of large bodies of water, including oceans, seas, lakes, and rivers (Figure 2.5). Surface waves are generated due to energy transfer within the water medium, leading to rhythmic oscillations that propagate across the water surface. Water waves exhibit a wide range of sizes, spanning from subtle undulations gently caressing the shoreline to colossal surges forcefully colliding with cliffs. The phenomenon of water wave formation is primarily influenced by wind, which serves as a mechanism for transferring energy to the water's surface. These waves are of utmost importance in the formation of coastlines, the erosion of shorelines, and the redistribution of sediments. Moreover, these entities support diverse marine ecosystems and serve as a preferred recreational destination for surfers and beach enthusiasts. A comprehensive comprehension of water waves holds significant importance in coastal engineering, oceanography, and the holistic admiration of the dynamic aesthetics exhibited by the expansive water bodies on our planet (Jeffreys, 1925).

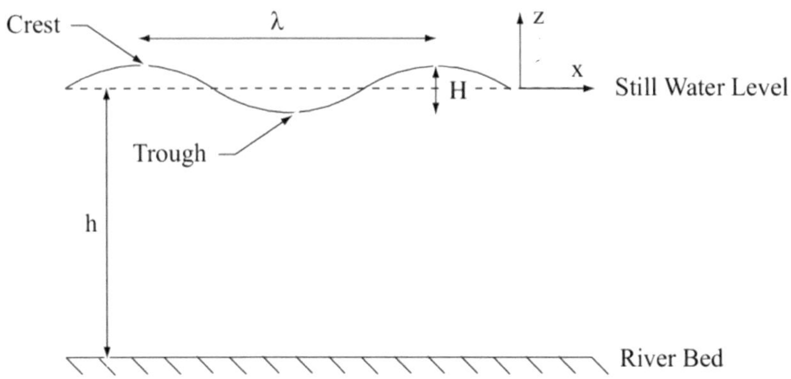

Figure 2.5. Illustration of a water wave (Source: David F Hill, Creative Commons License)

2.3.1. Ocean Waves

The captivating visual display of ocean waves' rhythmic undulations holds considerable importance in forming coastlines, exerting influence over weather patterns, and making substantial contributions to regulating Earth's climate. This section delves into the processes involved in generating and transmitting ocean waves, alongside the profound consequences that tsunamis inflict upon coastal regions (Tromans et al., 1991).

2.3.1.1. Formation and Propagation of Ocean Waves

Ocean waves are predominantly generated through the process of energy transmission from the wind to the surface of the water. The generation of ripples in the ocean is attributed to the frictional interaction between the moving air and the water, occurring as the wind traverses the expansive surface of the sea. The ripples undergo a process of growth and transformation, ultimately developing into fully-fledged waves due to the persistent influence of wind on the surface of the water (Onorato et al., 2009).

The dimensions and magnitude of oceanic waves are contingent upon various factors, encompassing the velocity, duration, and fetch (i.e., the span over which the wind persists) of the prevailing winds. Increased wind intensity and extended durations lead to the formation of larger waves characterized by higher energy levels. The fetch plays a crucial role in determining the extent to which wind can interact with unobstructed water, thereby exerting an influence on the size and intensity of waves (Stopa et al., 2018). Once generated, ocean waves spread across the expanse of the water's surface, effectively transmitting energy over considerable distances. The propagation of wave motion penetrates the water medium, although the magnitude of water displacement is comparatively minor about the total wave amplitude. As waves propagate from their source, they can traverse vast distances, spanning thousands of kilometers, ultimately arriving at remote shorelines (Mukhlas et al., 2022).

2.3.1.2. Tsunamis and their Impact on Coastal Areas

Tsunamis, commonly known as "seismic sea waves," are a distinct category of oceanic waves with potentially catastrophic outcomes. In contrast to wind-generated waves, tsunamis are triggered by abrupt and substantial ocean floor movements, commonly caused by subaquatic seismic activities, volcanic eruptions, or landslides. When an underwater seismic event transpires, it initiates substantial vertical displacements, displacing significant quantities of water. The release of energy during these events induces the elevation of the entire water column above the seafloor, forming a sequence of waves capable of propagating across vast ocean basins (Sonak et al., 2008).

Tsunamis are characterized by their significantly elongated wavelengths, frequently spanning hundreds of kilometers. However, their wave height remains relatively diminished when propagating through deep water. Nevertheless, when the tsunami nears coastal areas with less water depth, the waves undergo compression, significantly increasing their height. When a tsunami makes landfall, it has the potential to cause catastrophic consequences. The substantial magnitude of wave energy has the potential to submerge coastal regions situated at lower elevations, resulting in the occurrence of floods, extensive damage to infrastructure, and the unfortunate loss of human lives. Furthermore, it should be noted that tsunamis have the potential to manifest not only as a solitary wave but also as a sequence of consecutive waves, thereby exacerbating the extent of destruction inflicted (Szczuciński et al., 2006). The devastating impact of tsunamis was tragically exemplified in the Indian Ocean earthquake and tsunami of 2004, resulting in the loss of numerous lives in multiple nations along the Indian Ocean. Efforts aimed at reducing the impact of tsunamis on coastal communities encompass the implementation of early warning systems, enhanced comprehension of coastal geography to identify areas at risk, and the execution of public awareness campaigns to educate individuals on the requisite evacuation protocols in the event of a tsunami warning (Latief & Hadi, 2007).

2.3.2. Ripples and Waves in Lakes and Ponds

Lakes and ponds offer a tranquil environment where one can observe the delicate interplay of ripples and waves. These bodies of water are susceptible to the appealing forces of nature that produce surface disturbances. This section delves into the phenomenon of ripple and wave formation in lakes and ponds, specifically emphasizing the role of surface tension and capillary waves. Additionally, we examine the intricate interplay between these water movements and external forces such as wind and gravity (Aspler 1994).

2.3.2.1. Surface Tension and Capillary Waves

Surface tension is an essential phenomenon that occurs at the water interface, resulting from the cohesive interactions among water molecules. The phenomenon of cohesive behavior leads to a heightened attraction between water molecules at the surface, surpassing their attraction to the air above. Consequently, this gives rise to a "skin-like" effect that effectively hinders the infiltration of external forces. The relationship between surface tension and the energy required to disrupt the water surface is such that as surface tension increases, the amount of energy needed to interfere with the water surface also increases (Eisenmenger, 1959).

Capillary waves are small-amplitude waves that emerge on the surface of the water due to the equilibrium between the upward force generated by surface tension and the downward gravitational force. The amplitudes of these waves are generally smaller than one centimeter, while their wavelengths range from a few centimeters. Capillary waves are commonly observed on tranquil water surfaces, such as small ponds, where the influence of wind forces is negligible. Capillary waves are generated on the water's surface when it is disturbed by an object, such as a pebble or a gentle gust of wind. The water molecules near the disturbance undergo a transient augmentation in surface tension, resulting in a slight upward displacement. The vertical displacement in an upward direction gives rise to a minor crest, and subsequent restoration of the molecules to their initial positions leads to the formation of a trough behind the crest, resulting in a sequence of ripples (MacDowell et al., 2014).

2.3.2.2. Interactions with Wind and Gravity

The tranquility of capillary waves in lakes and ponds is frequently disrupted by the impact of wind due to their exposure to the natural environment.

The presence of wind induces pressure on the surface of the water, leading to the amplification of disturbances and the subsequent formation of more prominent waves. Gravity waves arise from the interaction between wind and the water's surface. These waves are distinguished by their greater amplitudes and longer wavelengths than capillary waves. The magnitude and amplitude of gravitational waves are subject to the influence of wind velocity, duration, and the extent of its propagation, commonly known as the fetch. The presence of powerful winds over an extensive stretch of water results in the generation of gravity waves of greater magnitude, which have the ability to propagate across vast distances.

Gravity waves become evident as the water surface of a lake or pond undergoes oscillations, resulting in periodic elevations and depressions. As the propagation of waves occurs within a water medium, energy is transferred, while the water molecules predominantly maintain their positions, merely oscillating in reaction to the advancing wave. The energy transmission process facilitates the propagation of gravity waves over extensive distances, resulting in the captivating phenomenon of undulating waves observed on expansive bodies of water. The dynamic and ever-changing water surface results from the interaction between capillary waves, which are produced by surface tension, and gravity waves, which are generated by wind. Under tranquil atmospheric conditions, capillary waves exhibit a gentle undulating motion on the water's surface. Conversely, during periods of strong winds, the prevailing presence is larger gravity waves, which assert their dominance over the water's surface (Phillips, 1985).

2.4. SEISMIC WAVES

Seismology is an academic discipline that systematically investigates seismic waves and their characteristics as they propagate through the Earth (Figure 2.6). Seismic events, such as earthquakes and volcanic activity, generate waves that offer valuable insights into the structure and composition of the Earth's interior. This section focuses on the fundamental aspects of seismology, specifically examining the structural composition of the Earth and the various classifications of seismic waves, such as P-waves and S-waves (Kuznetsov, 2011).

2.4.1. The Earth's Structure and Composition

The Earth comprises discrete layers, each possessing distinct and unique characteristics. Seismologists have employed seismic waves to enhance their comprehension of these layers and the constituent materials they consist of. The Earth can be classified into three primary layers:

- *Crust*: The crust refers to the Earth's outermost layer. The layer in question exhibits a relatively reduced thickness in comparison to the surrounding layers and is primarily comprised of compacted rock material. The thickness of the crust exhibits variation, wherein the oceanic crust is comparatively thinner and denser in comparison to the continental crust (Nestola et al., 2011).
- *Mantle*: The mantle is located beneath the Earth's crust and has a depth of approximately 2,900 kilometers (1,800 miles). The semi-solid layer under consideration comprises rock material that undergoes plastic deformation over extended periods of geological time. The mantle can be further subdivided into two distinct regions: the upper and lower mantle.
- *Core*: The core of the Earth is the innermost layer, comprising two separate regions known as the outer core and the inner core. The outer core is primarily comprised of a fluid mixture of iron and nickel. In contrast, the inner core exists as a solid sphere consisting predominantly of iron and nickel due to the substantial pressure exerted at its central region.

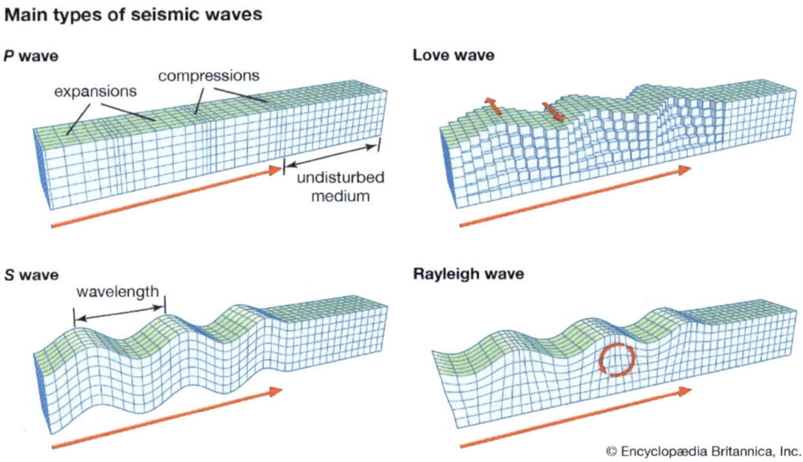

Figure 2.6. Schematic of the main types of the seismic waves (Source: Britannica, Creative Commons License).

Seismic waves are of utmost importance in mapping the various layers within the Earth's interior and discerning the boundaries that separate them.

Seismic waves undergo refraction and reflection at the boundaries between distinct layers within the Earth, thereby furnishing seismologists with significant insights into our planet's composition and density fluctuations (Kuwayama et al., 2008).

2.4.2. Types of Seismic Waves (e.g., P-Waves, S-Waves)

Seismic waves can be classified into two primary categories: body and surface waves. Body waves spread through the Earth's subsurface layers, whereas surface waves traverse along the Earth's outermost layer. The two main categories of body waves encompass P-waves, also known as primary waves, and S-waves also referred to as secondary waves.

- *P waves*: P-waves, also known as primary waves, exhibit the highest velocity among seismic waves and are the initial waves detected by seismographs in an earthquake. Seismic waves are compressional waves that propagate through the Earth by inducing parallel oscillations of particles in the direction of wave propagation. P-waves can propagate through both solid and liquid substances, enabling their transmission across the entirety of the Earth, encompassing its core.
- *S waves*: Secondary waves, also known as S-waves, are observed to follow primary waves, or P-waves, in the chronological order of seismic wave arrival. Shear waves are a type of wave that propagates by inducing particles to oscillate in a direction perpendicular to the wave's travel direction. In contrast to P-waves, S-waves cannot propagate through liquids, thereby precluding their transmission through the Earth's outer core. The identification of shadow zones, characterized by the absence of S-waves in specific Earth regions, served as initial empirical support for the presence of Earth's core.

In contrast, surface waves are primarily accountable for the harmful consequences of an earthquake near its epicenter. The motion of these waves is intricate, encompassing both horizontal and vertical components, and typically exhibits slower movement than body waves.

Seismologists employ the recorded arrival times of P-waves and S-waves, as detected by seismographs, to ascertain the epicenter and magnitude of an earthquake. By examining seismic wave properties, seismologists can acquire valuable knowledge about the origin, location, and depth of earthquakes, as well as the geological formations that these waves have traversed

(Sleep & Nakata, 2015).

2.4.3. Earthquakes and their Causes

Earthquakes are geological events that occur due to the abrupt release of energy within the Earth's crust. Geological phenomena possess the potential to exert substantial influences on both human civilizations and the natural environment. This section delves into the underlying factors contributing to earthquakes, specifically emphasizing faults and movements of tectonic plates. Additionally, it examines the impact of seismic waves on various structures (Polet & Kanamori, 2000).

Tip

Understanding the causes of earthquakes is essential for studying seismic waves. Tectonic plate movements, volcanic activity, and human-induced seismicity are some of the primary causes of earthquakes.

2.4.3.1. Faults and Tectonic Plate Movements

Earthquakes predominantly transpire along faults, fractures or regions of vulnerability in the Earth's crust where the tectonic plates interact. The lithosphere of the Earth is partitioned into multiple expansive and inflexible plates, which rest atop the partially fluid asthenosphere beneath them. The displacement and mutual influence of these lithospheric plates are accountable for the majority of seismic events occurring globally.

- *Transform boundaries*: Transform boundaries are a type of tectonic plate boundary where two plates slide past each other horizontally. This movement can occur in various directions, such as parallel or perpendicular. At transform boundaries, tectonic plates undergo horizontal sliding motion relative to each other. The inter-plate friction impedes the motion of the plates until the applied stress surpasses the resistance, leading to abrupt sliding along the fault plane. The abrupt discharge of energy results in seismic events known as earthquakes. The San Andreas Fault in California is widely recognized as an exemplar of a transform boundary, wherein the Pacific Plate and the North American Plate are engaged in lateral displacement relative to one another (Hanks, 1977).
- *Convergent boundaries*: They involve the collision of tectonic

plates, resulting in either subduction or crustal compression. The geological process known as subduction takes place when a tectonic plate is compelled to move beneath another plate and descend into the Earth's mantle. Powerful megathrust earthquakes can be caused by the significant pressure and friction experienced at the subduction zone. An instance that exemplifies this phenomenon is the seismic event that occurred in the Indian Ocean in 2004, which can be attributed to the subduction of the Indian Plate beneath the Burma Plate.

- *Divergent boundaries*: Divergent boundaries involve the separation of tectonic plates as they move away from each other. During the process of plate divergence, the movement of tectonic plates leads to the separation of the lithospheric plates. Consequently, molten rock, known as magma, ascends from the Earth's mantle to occupy the resulting void, thereby facilitating the formation of fresh crust. The level of volcanic and seismic activity observed at divergent boundaries is typically less intense in comparison to convergent boundaries. However, it is important to note that earthquakes do still occur as a result of the plates undergoing adjustments to accommodate the newly generated tensional stresses (Molnar, 1979).

2.4.3.2. Seismic Wave Effects on Structures

When an earthquake transpires, seismic waves disperse outward from the epicenter, which designates the point of origin. The transmission of energy through the Earth's crust by these waves can have catastrophic consequences on structures, especially in regions with high population density.

- *Surface waves*: Known for their slow velocity and high destructive potential, they represent the most detrimental form of seismic waves. Seismic waves propagate across the Earth's surface, inducing a rhythmic or oscillatory displacement of the ground. Surface waves are the primary agents behind the formidable vibrations that can lead to the structural failure of buildings, bridges, and various other constructions in an earthquake.
- *P-waves and S-waves*: P-waves and S-waves are additional factors contributing to the seismic vibrations experienced by structures during an earthquake. P-waves are characterized by their relatively high velocity and minimal impact on structural

integrity. In contrast, S-waves possess the potential to induce substantial shaking, thereby increasing the risk of structural failure, particularly in buildings that are inadequately designed or constructed (Semblat et al., 2008).

Structural engineers employ diverse methodologies to alleviate the impact of seismic waves on buildings and infrastructure. Several strategies can be employed to enhance the structural integrity of buildings in seismic-prone areas. These measures encompass using steel and concrete to fortify the construction, implementing base isolators to effectively absorb and dissipate seismic energy, and adhering to rigorous building codes and regulations specifically designed for earthquake-prone regions.

Implementing earthquake preparedness initiatives and educational programs is crucial in mitigating the adverse effects of seismic events on local communities. Early warning systems, earthquake drills, and public awareness campaigns educate individuals about the appropriate protocols to follow in the event of an earthquake, namely the "Drop, Cover, and Hold On" technique (Luco & Wong, 1982).

2.5. ATMOSPHERIC WAVES

Atmospheric waves are a subject of great interest and complexity within the Earth's atmosphere. A diverse array of wave types is encompassed, which includes sound waves, gravity waves, and atmospheric tides. Sound waves refer to pressure waves that can propagate through air, enabling sound perception and facilitating communication across varying distances. In contrast, gravity waves are vertical oscillations that can arise from various factors, including orographic lifting and convective processes. These waves have the potential to exert an impact on weather patterns and the general circulation of the atmosphere. Atmospheric tides refer to fluctuations in the temperature and pressure of the atmosphere that occur over extended periods. These oscillations are primarily influenced by the gravitational forces exerted by the Sun and the Moon. Examining atmospheric waves is of utmost importance in comprehending meteorological patterns, fluctuations in climate, and the dynamics of the constantly evolving atmosphere of our planet (Rossby, 1945).

2.5.1. Sound Waves in the Atmosphere

The atmosphere is not merely an inert space but a dynamic medium in which sound waves interact and propagate. Sound waves are of utmost importance in shaping our atmospheric encounters, ranging from the tranquil rustling of leaves to the magnificent rumble of thunder. This section delves into the characteristics of sound propagation in the air and the intriguing acoustic phenomena, such as thunder, that manifest within the atmosphere.

Remember

Sound waves in the atmosphere play a significant role in auditory communication. They travel through air and other gaseous mediums, allowing us to hear sounds from distant sources.

2.5.1.1. The Nature of Sound Propagation in Air

Sound is classified as a mechanical wave, necessitating a medium for its transmission. In the Earth's atmosphere context, air functions as the principal medium through which sound propagates. When an object undergoes vibration, it induces variations in the pressure of the adjacent air molecules. The variations above result in compressions and rarefactions, leading to the oscillation of air between areas characterized by elevated and diminished pressure.

The sound velocity in the atmosphere is subject to various factors, including temperature and humidity. In general, it is observed that the speed of sound is higher in air with higher temperatures. This phenomenon can be attributed to the increased energy of air molecules, which enables them to propagate vibrations more efficiently. The speed of sound is influenced by humidity, as moist air conducts sound at a slightly slower pace compared to dry air. This can be attributed to the presence of water vapor, which introduces additional molecular interactions.

The propagation of sound waves in the atmosphere can be influenced by a range of atmospheric phenomena, including temperature inversions and wind gradients. Temperature inversions are atmospheric phenomena characterized by a deviation from the typical decrease in temperature with increasing altitude, wherein the temperature increases with altitude. This phenomenon can give rise to the refraction of sound waves towards the Earth's surface, leading to sound propagation over abnormally extended distances, occasionally resulting in unforeseen acoustic phenomena (Kino & Ueno, 2008).

2.5.1.2. Thunder and Acoustic Phenomena

Thunder is an intriguing auditory phenomenon that manifests itself during thunderstorms. The phenomenon occurs due to the swift expansion of air caused by the high temperatures produced by a lightning discharge. When an electrical discharge known as lightning traverses through the atmosphere, it rapidly elevates the temperature of the adjacent air molecules to levels surpassing 27,000 degrees Celsius (48,632 degrees Fahrenheit) within an extremely brief timeframe. When subjected to high temperatures, the air undergoes a rapid expansion resulting in an explosive release of energy. This energy release generates a shockwave that travels through the surrounding atmosphere in the form of sound waves.

The auditory perception of thunder is subject to variability, which can be attributed to several factors, including the proximity of the lightning discharge, the prevailing temperature and humidity levels in the surrounding air, and the prevailing atmospheric conditions. The low-frequency rumble of distant thunder can be attributed to the scattering of sound waves as they propagate through the atmosphere. This scattering process leads to the preferential absorption of higher-frequency components, resulting in the perception of a low-frequency rumble. Consequently, the audible spectrum is predominantly characterized by low-frequency sounds. Certain instances of thunder can generate distinctive acoustic phenomena, such as the occurrence of "thunder rolls" and the manifestation of "echoes." Thunder rolls are a phenomenon that arises from the overlapping and reinforcement of sound waves generated by multiple lightning strikes, resulting in an extended rumbling auditory experience. Echoes may arise as a result of the phenomenon wherein sound waves encounter nearby objects, subsequently rebounding toward the observer and engendering the perceptual impression of multiple thunderclaps (Spivak et al., 2018).

In addition to thunder, various atmospheric phenomena have the capacity to generate captivating auditory experiences. The auditory phenomena of whistling winds, cracking ice, and the snapping of branches in strong winds are illustrative instances of atmospheric sounds that contribute to the diverse array of acoustic encounters within the realm of the natural environment. In summary, sound waves play a crucial role in the atmospheric milieu, influencing our perception of sound and offering valuable insights into atmospheric dynamics. Thunder is an impressive auditory phenomenon that occurs as a consequence of the intense heat generated by lightning, captivating and occasionally surprising individuals amidst thunderstorms.

Gaining knowledge about the transmission of sound waves in the atmosphere and the associated acoustic phenomena enhances our comprehension of the complex and dynamic characteristics of the surrounding air (Rahmani et al., 2020).

2.5.2. Atmospheric Gravity Waves

Atmospheric gravity waves are a captivating phenomenon in Earth's atmosphere (Figure 2.7). These waves can be classified as mechanical waves that propagate in a vertical direction, primarily influenced by the force of gravity. Various origins, such as orographic lifting, convection, and frontal boundaries, can be attributed to them. When air is compelled to ascend or descend due to these mechanisms, it generates undulating disturbances that propagate vertically within the atmosphere. Atmospheric gravity waves have the potential to manifest at various altitudes, ranging from the Earth's surface boundary layer to the higher regions of the atmosphere. They have a notable impact on the formation of weather patterns, the adjustment of jet streams' intensity and location, and the alteration of broad-scale atmospheric circulation patterns. Moreover, they can potentially influence the process of cloud formation and play a role in climate feedback mechanisms. A comprehensive comprehension of atmospheric gravity waves holds significant importance in advancing weather forecasting models and deepening our knowledge regarding the dynamic characteristics of the Earth's atmosphere (Hocke & Schlegel, 1996).

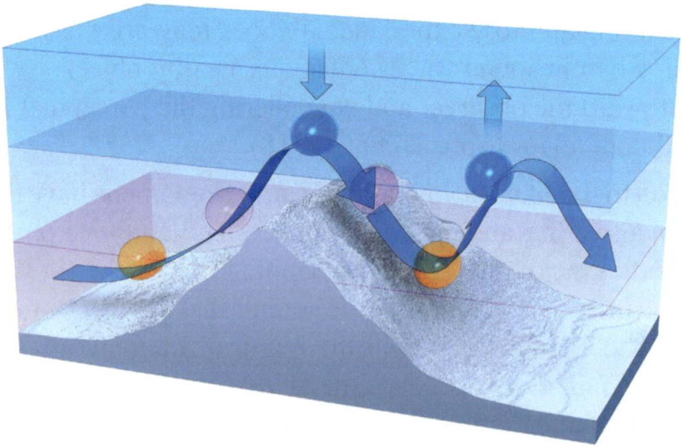

Figure 2.7. Representation of the atmospheric gravity waves (Source: National Weather Service, Creative Commons License)

2.5.2.1. Formation and Characteristics of Gravity Waves

Atmospheric gravity waves are intriguing and indispensable phenomena within Earth's atmosphere. The observed waves differ from the gravitational waves postulated by Einstein's theory of general relativity. Rather, they represent a distinct class of mechanical waves that propagate vertically within the atmosphere due to gravitational force. This inquiry aims to investigate the genesis and attributes of atmospheric gravity waves.

- *Formation*: Various mechanisms, such as orographic lifting, convection, and frontal boundaries, can generate atmospheric gravity waves. Orographic gravity waves arise due to the upward displacement of air due to the obstruction posed by mountains or other notable topographic elements. As the vertical motion of air ascends, it undergoes expansion and subsequently cools, resulting in undulating disturbances within the atmosphere. Convection-induced gravity waves arise due to warm air's upward displacement and cooler air's subsequent downward movement within convective clouds. Gravity waves can also be generated by frontal boundaries, which occur when air masses of varying temperatures and densities interact (Bakhmet'eva et al., 2018).
- *Characteristics*: Atmospheric gravity waves possess distinct characteristics that set them apart from other atmospheric phenomena. In contrast to sound waves, which exhibit longitudinal wave properties, gravity waves are characterized by their transverse wave nature. This implies that the air particles exhibit motion that is perpendicular to the direction of wave propagation. Gravity waves exhibit a wide range of wavelengths, spanning from a few kilometers to several hundred kilometers, while their periods can vary from minutes to hours.

Gravity waves have the potential to manifest at varying elevations within the Earth's atmosphere. Low-level gravity waves are commonly observed within the boundary layer, which refers to the atmospheric layer closest to the Earth's surface. Gravity waves of intermediate and higher magnitudes propagate into the lowermost layer of the Earth's atmosphere, known as the troposphere, while those of greater intensity extend further into the subsequent layer called the stratosphere. An intriguing characteristic of gravity waves is their ability to transport momentum and energy within the atmosphere vertically. The upward propagation of waves has the potential to influence wind patterns and temperature distributions, thereby impacting the overall circulation of the atmosphere (Medvedev & Klaassen, 2000).

2.5.2.2. Impact on Weather and Climate

Atmospheric gravity waves are of significant importance in the dynamics of weather and climate. The impact of their presence spans various levels, ranging from the regional level, where they can generate turbulence and contribute to the formation of clouds, to the global level, where they can exert influence on extensive patterns of atmospheric circulation.

- *Local effects*: Gravity waves have the potential to induce turbulence, particularly within the lower atmosphere. When waves come into contact with fluctuations in temperature and wind, they have the potential to undergo a process known as breaking, which subsequently results in the intermingling of air masses and the emergence of turbulence. In regions characterized by elevated topography, the presence of orographic gravity waves can develop lenticular clouds, which manifest as immobile cloud formations resembling convex lenses positioned in the downwind direction relative to the mountains.
- *Cloud Formation*: The formation and patterns of clouds can be influenced by gravity waves. Cloud formation can occur when air ascends and undergoes cooling as a result of gravity waves, causing it to reach its dew point. In certain instances, the presence of gravity waves can give rise to orderly cloud formations referred to as "cloud streets," which exhibit a parallel alignment with the prevailing wind direction (Paraskevis et al., 2021).
- *Large-scale circulation*: The influence of atmospheric gravity waves on the atmospheric circulation patterns is observed globally. The ability to modulate the intensity and location of jet streams, influence the polar vortex dynamics, and contribute to the formation of atmospheric blocking patterns is within their purview. The aforementioned significant influences can potentially affect weather patterns and long-term climate variability.
- *Climate feedback*: Gravity waves are also involved in climate feedback mechanisms. Changes in the intensity and spatial distribution of gravity waves can impact the atmospheric transport of ozone and water vapor, thereby exerting an influence on the temperature and composition of the atmosphere.

A comprehensive comprehension of the dynamics exhibited by atmospheric gravity waves is imperative in order to enhance the accuracy and reliability of weather prediction models and climate simulations. The waves above are intricate phenomena that engage in interactions with various atmospheric processes, thereby playing a role in shaping the overall dynamic behavior of Earth's atmosphere (Poole et al., 2019).

2.6. ELECTROMAGNETIC WAVES IN NATURE

Electromagnetic waves in the natural environment exhibit a wide range of characteristics and hold a fundamental position in the realm of physics. These waves represent a type of energy that is transmitted through empty space, allowing them to travel through the vacuum found in outer space. The electromagnetic spectrum encompasses a broad array of wavelengths, spanning from gamma rays with high energy to radio waves with low energy. Several prevalent electromagnetic waves in the natural world include visible light, which enables human perception of the surrounding environment, and radio waves, which serve as a medium for wireless transmission and broadcasting. In addition, it should be noted that infrared waves facilitate the process of thermal imaging. In contrast, ultraviolet waves are of paramount importance in various biological processes such as photosynthesis and are also accountable for the occurrence of sunburns. X-rays and gamma rays are employed in medical imaging to visualize internal structures and diagnose various medical conditions. Additionally, these types of radiation can penetrate different materials, making them valuable tools for inspection and quality control purposes. The investigation of electromagnetic waves in the natural environment is of utmost importance in comprehending diverse natural phenomena, ranging from celestial entities' dynamics to biological systems' operations (Mani et al., 2002).

2.6.1. Sunlight and Solar Waves

Sunlight and solar waves are integral elements of the natural environment, originating from the celestial body in closest proximity to Earth, namely the Sun. Sunlight is a type of electromagnetic radiation that includes a wide range of wavelengths, spanning from ultraviolet to infrared rays. These electromagnetic waves propagate through the vacuum of space and ultimately reach the Earth's surface, where they play a crucial role in sustaining life and facilitating a range of atmospheric and climatic phenomena. The role of sunlight in photosynthesis is of utmost importance as it enables plants to convert carbon dioxide and water into carbohydrates rich in energy while simultaneously releasing oxygen. Solar waves encompass solar flares, which manifest as vigorous emissions of electromagnetic radiation, particles, and shockwaves originating from the Sun's outer layer. These flares can substantially impact space weather, potentially leading to disruptions in satellite communications and terrestrial power grids. Comprehending the properties of sunlight and solar waves is imperative in the pursuit of comprehensively investigating the intricate dynamics of our solar system, the Earth's climate, and the consequential ramifications of solar activity on our technological infrastructure (Mozer et al., 2021).

2.6.1.1. Solar Radiation and its Interaction with the Atmosphere

Sunlight, the radiant energy released by the Sun, is crucial in supporting life on Earth and facilitating diverse natural phenomena. This section examines the phenomenon of solar radiation and its intricate interplay with the Earth's atmosphere, exerting a profound influence on our surrounding environment and climate.

Solar radiation refers to the electromagnetic radiation emitted by the Sun, encompassing a wide range of wavelengths. It is a fundamental source of The Sun that releases energy through the emission of electromagnetic waves spanning a wide range of frequencies, encompassing high-energy gamma rays to low-energy radio waves. Nevertheless, it is important to note that the solar spectrum, which encompasses a limited portion of the electromagnetic spectrum, is the only fraction that can penetrate the Earth's surface. The solar spectrum encompasses ultraviolet (UV) radiation, visible light, and infrared (IR) radiation. The portion of the solar spectrum perceptible to the human eye is known as visible light, characterized by its vivid array of colors. The interaction between solar radiation and the Earth's atmosphere occurs when solar radiation encounters the different components of the atmosphere. Various atmospheric gases, such as ozone, water vapor, and carbon dioxide, absorb some of the solar radiation that reaches the Earth's atmosphere. The wavelengths of energy that are absorbed play a role in the heating of the atmosphere and the greenhouse effect, ultimately contributing to the maintenance of Earth's average temperature at a level suitable for sustaining life. A fraction of solar radiation is dispersed by atmospheric particles such as air molecules and aerosols. The phenomenon of scattering is accountable for the aesthetically pleasing coloration observed in the sky during the periods of sunrise and sunset. The phenomenon of the blue sky can be attributed to the efficient scattering of blue light, which possesses a shorter wavelength. Conversely, the prevalence of longer-wavelength red and orange hues can be observed during sunsets and sunrises (Bai et al., 2022).

2.6.1.2. Solar Flares and Space Weather Effects

In addition to serving as a source of vital energy for sustaining life, the Sun also displays dynamic and occasionally dramatic phenomena referred to as solar flares. The flares under consideration have the capacity to emit significant quantities of energy, encompassing electromagnetic radiation, particles, and shockwaves. These emissions have the potential to impact space weather and our technological infrastructure.

- ***Solar flares***: Solar flares are abrupt outbursts occurring on the surface of the Sun, resulting from the liberation of magnetic energy. When the magnetic fields in the atmosphere of the Sun become entangled and distorted, they can undergo a process known as rapid reconnection, releasing a significant amount of energy. The phenomenon described involves the rapid emission of energy, which leads to the acceleration of charged particles, the generation of high-energy X-rays and gamma rays, and the occurrence of intense bursts of electromagnetic radiation spanning the entire electromagnetic spectrum.
- ***Space Weather Effects***: The effects of space weather are primarily attributed to the electromagnetic radiation and charged particles emitted during solar flares. These phenomena have notable consequences on the space environment encompassing the Earth. As an illustration:
  - *Geomagnetic storms*: They can be initiated on Earth due to solar flares. The interaction between the solar wind, a flow of charged particles emitted by the Sun, and Earth's magnetic field can induce disruptions within the magnetosphere. This phenomenon has the potential to generate auroras, specifically the Northern and Southern Lights, and in severe instances, it can cause disturbances in power grids, satellite communications, and GPS systems.
  - *Solar Particle Events*: Solar Particle Events (SPEs) pose a significant risk to astronauts and satellites in space due to the release of high-energy charged particles during solar flares. During periods of heightened solar particle events, it may be necessary for astronauts situated in space or beyond Earth's protective atmosphere to seek refuge in order to mitigate the potential risks associated with detrimental radiation exposure.

– *Radio blackouts*: Radio blackouts can occur as a result of the emission of high-energy electromagnetic radiation during solar flares, leading to disruptions in radio communications, particularly at specific frequencies.

Solar flares possess the capacity to impact our technological infrastructure, yet they also play a role in the captivating phenomenon of auroras in polar regions. These auroras visually represent the intricate interplay between solar radiation and Earth's magnetic field (Yiğit et al., 2016).

2.6.2. Natural Sources of Radio Waves

The origins of radio waves in nature are extensive and captivating, arising from a multitude of celestial and terrestrial phenomena. Cosmic radio bursts, such as Fast Radio Bursts (FRBs), exhibit the emission of highly concentrated radio wave bursts originating from remote areas of the cosmos, thereby presenting a perplexing enigma to astronomers regarding their sources. Celestial entities, such as pulsars and active galactic nuclei, play a significant role in shaping the radio wave environment, providing valuable observations regarding the characteristics and dynamics of magnetized neutron stars and supermassive black holes.

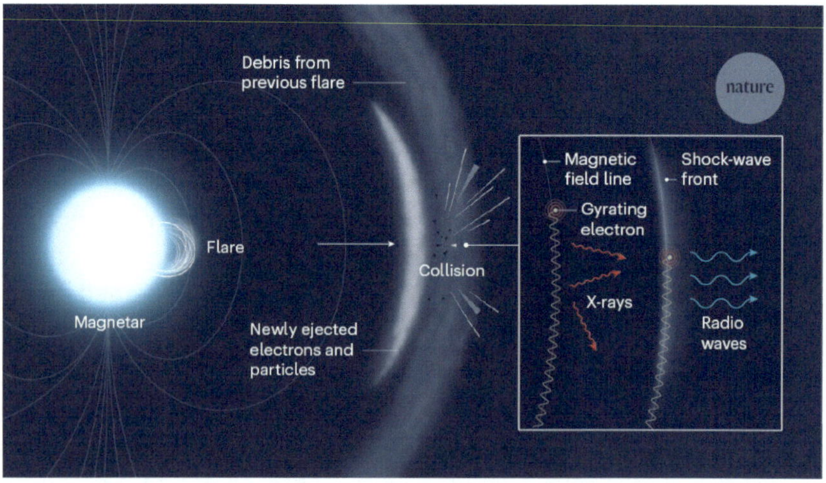

Figure 2.8. A fast radio burst in our own Galaxy (Source: Amanda Weltman, Creative Commons License).

In proximity, the Sun emits radio waves during occurrences such as solar flares and other solar events, thereby influencing space weather. Radio waves are generated by terrestrial phenomena, such as lightning discharges that occur during thunderstorms. The utilization of natural sources of radio waves enhances our comprehension of the cosmos and facilitates the investigation of enigmatic phenomena within the universe (Figure 2.8).

2.6.2.1. Cosmic Radio Bursts

Cosmic radio bursts are enigmatic and ephemeral phenomena that emanate powerful surges of radio waves from remote sectors of the cosmos. Astronomers have been captivated and intrigued by these potent radio signals since their initial detection. In this discourse, we shall investigate the characteristics and genesis of cosmic radio bursts.

- *Characteristics*: The detection and study of cosmic radio bursts pose significant challenges due to their extremely brief durations, often lasting only a fraction of a second. The initial detection of these signals occurred during the 1960s when radio astronomers detected anomalous signals that exhibited indications of extragalactic origins. The detected signals exhibited distinct characteristics inconsistent with any previously identified radio emissions originating from celestial entities during that period.
- *Fast Radio Bursts (FRBs)*: FRBs are the prevailing category of cosmic radio bursts. Fast radio bursts (FRBs) are distinguished by their swift and vigorous emission of radio waves spanning a wide spectrum of frequencies. The precise mechanism underlying fast radio bursts (FRBs) continues to be a topic of ongoing investigation and conjecture.
- *Origins*: The origins of fast radio bursts (FRBs) remain incomplete, although numerous hypotheses have been put forth. One potential explanation for these phenomena is their association with cataclysmic astrophysical occurrences, such as the fusion of neutron stars or the gravitational collapse of magnetized neutron stars resulting in the formation of black holes. An alternative hypothesis posits a potential association between Fast Radio Bursts (FRBs) and magnetars, characterized by their intense magnetic fields and ability to emit substantial quantities of energy in the radio wave spectrum.
- *Deep Space Probing*: The investigation of cosmic radio bursts has facilitated the exploration of the vast expanse of the universe, enabling astronomers to examine distant regions that extend well

beyond our galaxy. By examining the dispersion and scattering phenomena exhibited by radio waves during their propagation across intergalactic space, researchers can acquire significant insights into the spatial arrangement of matter and magnetic fields within the expansive realm of the universe (Macquart et al., 2020).

2.6.2.2. Radio Emissions from Celestial Bodies

Radio waves are produced by a multitude of celestial entities dispersed throughout the cosmos, thereby furnishing astronomers with significant insights into the underlying physical phenomena transpiring within these entities. The following are noteworthy sources of radio emissions:

- *Pulsars*: Pulsars are celestial objects known as highly magnetized neutron stars, which exhibit the emission of electromagnetic radiation, encompassing radio waves, originating from their magnetic poles. The emission of regular radio pulses occurs as a consequence of the rotational movement of the star, causing its beams to traverse through space akin to the behavior of a celestial lighthouse. The radio signals emitted by these pulses are observable on Earth, enabling astronomers to investigate the characteristics of pulsars and delve into the realm of fundamental physics.
- *Supernova remnants*: Supernova remnants are formed as a consequence of the explosive demise of massive stars, wherein a substantial amount of energy is discharged. Supernova remnants, which are the residual products of these explosive events, can emit radio waves due to their interaction with the encompassing interstellar medium. The examination of radio emissions originating from supernova remnants facilitates astronomers in comprehending the dynamic processes that influence the development of galaxies (Dagkesamanskii & Zheleznykh, 1989).
- *Galactic Center*: Located within the Milky Way galaxy, the galactic center exhibits notable emissions of radio signals that have been correlated with the existence of a supermassive black hole commonly referred to as Sagittarius A*. The aforementioned black hole exhibits the process of accretion, wherein it attracts matter from its immediate environment. This gravitational attraction causes the infalling material to undergo heating and

engage in interactions with magnetic fields, ultimately resulting in the emission of potent radio waves.

- *Extragalactic Radio Sources*: Radio galaxies and quasars, which are classified as active galactic nuclei, exhibit significant radio wave emission capabilities. The sources under consideration are derived from supermassive black holes located at the central regions of galaxies that are situated at a significant distance from us. These black holes are actively accreting matter, a process through which they consume surrounding material and subsequently emit substantial quantities of energy, including radio emissions.

Radio telescopes, which possess a high sensitivity to radio waves, have played a pivotal role in detecting and investigating diverse celestial radio sources. These instruments enable astronomers to conduct observations of the universe distinctively, thereby unveiling concealed phenomena and facilitating a comprehensive comprehension of the cosmos (Afraimovich, 1981).

2.7. BIOLOGICAL AND BIOLOGICAL-PHYSICAL WAVES

The domain of wave phenomena in the natural world is encompassed by biological and biological-physical waves, which elicit great fascination. The phenomenon of bioluminescence and its role in inter-organism communication highlights the remarkable capacity of living organisms to generate and release light to convey messages, protect themselves, and attract others. The utilization of sound-based communication in the animal kingdom, encompassing vocalizations and acoustic signals, serves as a fundamental mechanism for transmitting information about reproductive activities, providing cautionary alerts, and establishing and preserving social connections. Moreover, certain animal species exhibit the extraordinary ability of echolocation, employing sound waves for the purpose of navigation and prey detection, as demonstrated by bats and marine mammals. The examination of these waves within the biological realm yields important insights into the complicated communication and survival tactics employed by various organisms, thereby augmenting our understanding and admiration of the intricacy and variety of life on our planet (Chavez et al., 1998).

2.7.1. Bioluminescence and Light Communication

Bioluminescence is an intriguing natural occurrence wherein living organisms generate and release light. The remarkable capacity to produce light is crucial in diverse species' communication and ecological dynamics. Within this particular section, we delve into the intricate process of light production in various organisms and examine the ecological importance of bioluminescence.

Did You Know?

Bioluminescence, a fascinating natural phenomenon, involves the production and emission of light by living organisms. Some marine organisms, like fireflies and certain deep-sea creatures, use bioluminescence for communication, attracting mates, and defense against predators.

2.7.1.1. Light Production in Organisms

Bioluminescence is a phenomenon characterized by a chemical reaction occurring within specialized cells or organs of living organisms. The three fundamental constituents for this process encompass a luminescent molecule termed luciferin, an enzyme denoted as luciferase, and oxygen. When these constituents combine in the presence of oxygen, they undergo a chemical reaction that liberates energy in the form of light.

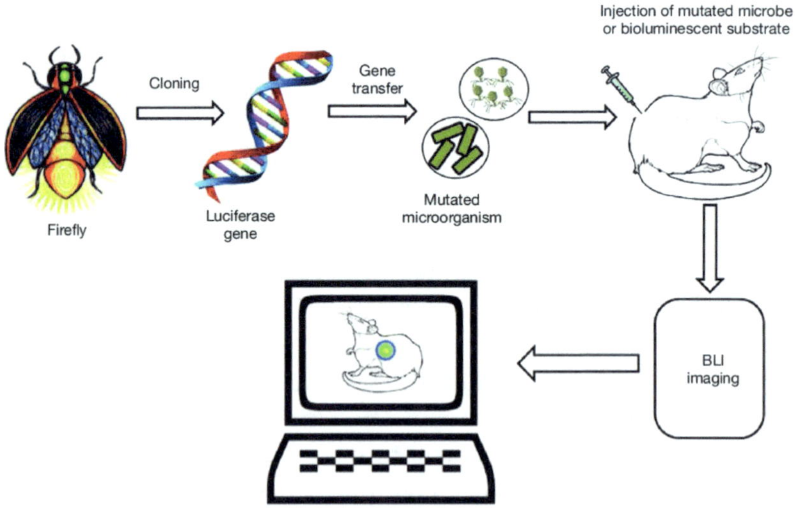

Figure 2.9. Schematic representation of bioluminescent imaging (Source: Pala Rajasekharreddy, Creative Commons License)

A variety of organisms employ bioluminescence for diverse functions. For example, certain marine plankton species, such as dinoflagellates, employ bioluminescence as self-defense. When subjected to disturbance, plankton exhibit transient bioluminescent emissions, which have the potential to elicit surprise or disorientation in potential predators, thereby allowing the plankton to evade capture. In certain instances, Bioluminescence serves various purposes, including communication, mate attraction, and prey attraction.

An exemplary illustration of bioluminescence is observed in fireflies. Male fireflies utilize their bioluminescent flashing light patterns to attract females. Every individual firefly species possesses a distinct flashing pattern, which serves as a visual cue to recognize and attract potential mates belonging to the same species. When a male firefly perceives a female firefly exhibiting a flashing pattern that is compatible with his own, he exhibits a distinct sequence of flashes in response, thereby engaging in a reciprocal interaction until the mating process is successfully completed (Gattuso et al., 2006).

2.7.1.2. Ecological Significance of Bioluminescence

Bioluminescence plays a pivotal ecological function in diverse marine ecosystems.

- Within the expansive obscurity of the ocean's profound depths, bioluminescent organisms orchestrate a captivating exhibition of luminosity that possesses both aesthetic allure and useful purpose.
- Certain species of squids and fish dwelling in the deep sea possess bioluminescent organs known as photophores, which serve as a form of camouflage. The photophores possess the ability to produce luminescence that is consistent with the prevailing light conditions in their surroundings. This adaptation enables the organisms to effectively camouflage themselves, thereby minimizing the risk of being detected by both predators and potential prey.
- Bioluminescence is frequently employed as a mechanism for mate attraction in various organisms. In numerous marine species, both males and females engage in bioluminescent displays as a means to locate appropriate mates for the purpose of reproduction. The luminous displays function as courtship signals, facilitating the identification of conspecifics during the reproductive period.

- Prey attraction is observed in predatory species, including specific deep-sea fish and squids, which employ bioluminescence as a mechanism to lure and capture their prey effectively. These organisms produce bioluminescent flashes in order to attract smaller organisms, thereby facilitating the capture and consumption of their prey.
- Bioluminescence can also manifest as a mutualistic symbiotic relationship wherein two species benefit from their association. An illustration can be found in the symbiotic relationship between the Hawaiian bobtail squid and bioluminescent bacteria. The squid facilitates a mutually beneficial relationship with bacteria by offering a secure and nourishing habitat within its light organ. In reciprocation, the bacteria generate luminescence, aiding the squid in concealing itself from potential predators situated beneath it.
- Predation avoidance can be achieved by utilizing bioluminescence in specific marine organisms. Certain species can secrete bioluminescent fluid or expel luminous particles into their surrounding aquatic environment as a defensive mechanism. This behavior serves to perplex and divert the attention of potential predators, thereby creating an opportunity for the prey to evade capture (Cusick & Widder, 2020).

2.7.2. Sound and Communication in the Animal Kingdom

The auditory abilities and communicative behaviors observed in various animal species play a crucial role in their survival and social dynamics. A diverse range of animal species, including avian, mammalian, and insect taxa, employ vocalizations and acoustic signals to communicate information. Avian species employ melodious vocalizations as a means to delineate and defend their territories, as well as to allure potential mates. Conversely, canid mammals such as wolves and coyotes utilize howling to communicate to strengthen social connections within their respective packs. Insects are known to engage in acoustic communication, exemplified by the production of chirping sounds by crickets to attract potential mates. Moreover, specific animal species, such as bats and marine mammals, have evolved the ability to employ echolocation, an advanced mechanism of sound-based orientation, enabling them to perceive their surroundings through auditory cues and effectively locate their prey. The investigation of auditory phenomena and inter-species communication within the animal realm serves to expand our comprehension of their intricate behavioral patterns, thereby fostering a more profound connection to the diverse intricacies of life on our planet (Marler, 1967).

2.7.2.1. Animal Vocalizations and Acoustic Signals

The role of sound in communication among different animal species is of utmost importance. Animal vocalizations and acoustic signals serve crucial purposes such as mating, defending territories, alerting to potential threats, and fostering social cohesion, encompassing a wide range of sounds, including melodic bird songs and intricate underwater calls. This inquiry aims to examine the wide range and importance of sound-based communication within the animal kingdom (Figure 2.10).

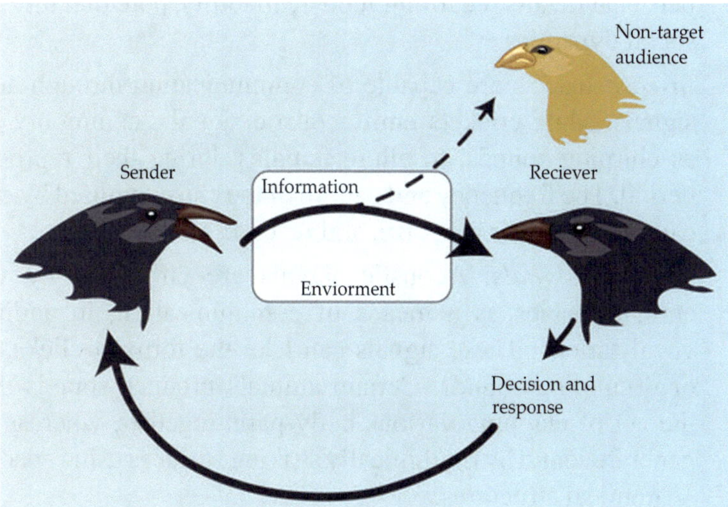

Figure 2.10. A model of animal communication (Source: Erin Gillam, Creative Commons License).

- *Vocalizations*: Vocalization refers to deliberate sounds emitted by animals with the purpose of communication. These vocalizations can manifest in diverse manners, encompassing calls, songs, chirps, howls, and roars. Various species have developed distinct vocalizations that are specifically adapted to their individual requirements (Wiley & Richards, 1978).
- *Birds*: Birds are widely recognized for their varied and intricate vocalizations. Songbirds employ melodious vocalizations as

a means to establish and delineate their territories, as well as to allure potential mates. The elaborate composition of these musical pieces serves as a means of conveying insights about the vocalist's physical well-being, vitality, and hereditary adaptability. Conversely, alarm calls serve the purpose of notifying fellow group members regarding the existence of predators, thereby enabling them to engage in evasive maneuvers.

- *Mammals*: Mammals employ vocalizations as a means of communication as well. Canids, such as wolves and coyotes, employ vocalizations known as howls to communicate their whereabouts to other pack members and strengthen social connections. Similarly, primates employ a diverse array of vocalizations, encompassing calls and hoots, to communicate pertinent details regarding food availability, potential threats, and social dynamics.
- *Insects*: Insects are capable of communicating through auditory signals. Male crickets emit acoustic signals, commonly known as chirping sounds, to allure females during their reproductive period. The frequency and pattern of the chirps emitted by crickets can indicate their age, size, and readiness for mating.
- *Acoustic signals*: Acoustic signals are employed by various animal species as a means of communication, in addition to vocalizations. These signals can take the form of clicks, snaps, or drumming sounds. Certain animals produce sounds through the act of clapping various body parts together, whereas others generate sound by rhythmically striking surfaces using specialized anatomical structures.
- *Frogs and toads*: Amphibians, namely frogs and toads, are widely recognized for their ability to engage in acoustic signaling. During the reproductive period, male individuals produce distinct vocalizations in order to attract female counterparts. Every species possesses a distinct vocalization, enabling individuals to identify and distinguish members of their own species for the purpose of reproduction.
- *Insects and arachnids*: Insects and arachnids employ acoustic signals as a component of their courtship rituals. Male crickets emit auditory signals in the form of chirping sounds with the purpose of attracting females. Conversely, male grasshoppers

and katydids generate distinctive calling songs (Burchardt & Knörnschild, 2020).

2.7.2.2. Echolocation in Bats and Marine Mammals

Echolocation is an extraordinary adaptation employed by specific animal taxa, notably bats and marine mammals, to effectively traverse their surroundings and detect and capture prey. The process of echolocation involves the emission of sound waves and the subsequent reception of their echoes, which are then utilized to construct a cognitive representation of the surrounding environment. These animals can utilize echolocation, allowing them to perceive their surroundings through sound, particularly in conditions of reduced illumination or complete darkness.

Bats: Bats are widely recognized as one of the most prominent examples of animals that utilize echolocation. These organisms produce high-frequency sound pulses from their oral or nasal cavities and perceive the resulting echoes that are reflected upon encountering obstacles in their trajectory. Through the analysis of echo timing and intensity, bats can accurately ascertain the position, dimensions, and motion of their prey, even under conditions of total absence of light (Brinkløv et al., 2022).

Marine mammals: Marine mammals, including dolphins, whales, and certain seal species, have developed the capacity for echolocation. These animals produce auditory signals resembling clicks or sonar by utilizing specialized anatomical structures, such as the melon in dolphins or the nasal passages in whales. Marine mammals utilize the reverberations emitted by these signals to facilitate their navigation, obstacle detection, and prey localization within the extensive oceanic environments.

Echolocation in Action: Echolocation can be observed when a bat or marine mammal emits a sound pulse, which then propagates through the surrounding medium, be it air or water, until it interacts with an object. The sound wave subsequently undergoes reflection, resulting in the formation of an echo, which is then detected by the animal's exceptionally perceptive auditory organs. The temporal interval between the emission and reception of the echo, in conjunction with the alteration in frequency resulting from the object's motion, yields significant insights into the object's spatial proximity and kinematics.

Bats and marine mammals possess an exceptional level of precision in their echolocation abilities, enabling them to detect and accurately locate prey within intricate and obstructed surroundings. As an illustration, dolphins possess the ability to locate diminutive fish concealed within the sand or mud on the ocean floor, whereas bats exhibit the capacity to seize airborne insects in flight. Sound-based communication exhibits a wide range of prevalence and diversity throughout the animal kingdom. Vocalizations and acoustic signals play pivotal roles in communicating information about reproductive behavior, social interactions, and potential threats. Echolocation is a highly specialized mechanism of sound-based navigation employed by bats and marine mammals to perceive their environment and accurately locate prey effectively. Examining auditory phenomena and communicative behaviors within the animal realm provides intriguing perspectives on the multifaceted and intricate mechanisms through which animals engage and adapt to their surroundings (Evans & Awbrey, 1988).

2.8. FUTURE RESEARCH IN UNDERSTANDING NATURAL WAVES

Investigating natural waves is an intriguing and continuously developing area of scholarly inquiry that holds significant implications for diverse scientific domains and practical implementations. Advancements in the understanding of natural waves, ranging from the behavior of ocean waves to the investigation of cosmic radio bursts, offer potential solutions to significant challenges and contribute to expanding our knowledge about the universe. This article aims to examine prospective avenues for future research and potential applications in the realm of comprehending natural wave phenomena (Schöbel et al., 2023).

- *Ocean waves*: The examination of ocean waves is a significant field of study, particularly in light of the increasing apprehension regarding climate change and its consequences on the rise of sea levels and the erosion of coastlines. Future research endeavors aim to enhance our capacity in modeling and forecasting extreme wave occurrences, such as storm surges and tsunamis, to improve coastal planning and disaster preparedness. Moreover, the progress made in ocean wave research has the potential to facilitate the advancement of renewable energy technologies, specifically wave energy converters. These devices are designed to capture and utilize the kinetic energy in ocean waves to produce electricity.
- *Seismic waves*: Research on seismic waves is constantly progressing in order to enhance earthquake early warning

systems and assess seismic hazards more effectively. By gaining a deeper comprehension of the transmission and dynamics of seismic waves within the Earth's crust, researchers can furnish more precise and prompt alerts regarding imminent earthquakes. This enhanced knowledge empowers communities to adopt proactive measures to mitigate harm and preserve human lives. Furthermore, the continuous investigation in this field contributes to enhancing building codes and engineering methodologies to develop structures that exhibit increased resilience against seismic events (Frölicher & Laufkötter, 2018).

- *Atmospheric Waves*: The study of atmospheric waves provides useful knowledge of climate variability and its potential influence on weather patterns. Engaging in advanced research on atmospheric gravity waves and their interactions with climate systems holds the potential to enhance the accuracy of climate models and advance the precision of long-term weather forecasting. Furthermore, the investigation of sound waves in the atmosphere exhibits potential for utilization in atmospheric remote sensing, facilitating the examination of temperature, pressure, and wind patterns in the higher regions of the atmosphere by scientists (Smith, 2011).
- *Electromagnetic Waves*: Future research on electromagnetic waves encompasses a diverse array of applications, spanning from the investigation of cosmic radio bursts and the elucidation of fast radio burst origins to the examination of celestial bodies' properties through the lens of radio astronomy. (Werner et al., 2013).
- *Matter waves*: Matter waves are of great importance in quantum mechanics as they offer valuable insights into the behavior exhibited by subatomic particles. The forthcoming investigations in this domain aspire to expand the frontiers of quantum technology, encompassing quantum computing and quantum communication. Utilizing the distinct characteristics of matter waves has the potential to bring about a significant transformation in information processing, cryptography, and secure communication. This advancement presents viable solutions to challenges that currently exceed the capacities of classical computing systems (Masten, 2007).
- *Biological and Biological-Physical Waves*: Investigating waves

within the biological realm presents intriguing prospects for diverse applications. The study of bioluminescence and light communication has the potential to stimulate progress in bio-inspired lighting technologies and make valuable contributions to medical imaging and diagnostics by creating bioluminescent markers. The comprehension of sound-based communication within the animal kingdom can also contribute to advancing bioacoustic research, thereby facilitating the development of innovative methodologies for tracking wildlife and conservation endeavors (Burke & Crocker, 2020).

- *Solar Waves*: The study of solar waves continues to be of utmost importance in the field of space weather forecasting and in enhancing our comprehension of the Sun's impact on Earth's climate. The progression in this particular domain has the potential to enhance the precision of space weather predictions, thereby facilitating enhanced safeguarding measures for satellites, power grids, and other vital infrastructure against solar storms and geomagnetic disturbances.
- *Cross-Disciplinary Research*: Cross-disciplinary research is considered to be the key to advancing our understanding of natural waves. This approach emphasizes the importance of fostering collaboration among different scientific disciplines. The integration of knowledge from various disciplines, such as physics, biology, geology, astronomy, and engineering, through cross-disciplinary research has the potential to yield significant breakthroughs and foster the development of novel applications. One example of potential synergy lies in the integration of ocean wave research with renewable energy technologies, which holds promise for enhancing the efficiency of wave energy converters. This collaborative approach has the potential to contribute significantly to the development of sustainable energy solutions (Mach et al., 2020).

Activity 2.1

You are a geologist conducting research on earthquakes for a renowned geological institute. While studying seismic waves, you come across a seismogram showing a distinct pattern of P-waves and S-waves during an earthquake event. How would you interpret the data from the seismogram to determine the earthquake's epicenter and magnitude? Additionally, how can this information contribute to improving earthquake preparedness and safety in vulnerable regions?

SUMMARY

The chapter titled "Wave Phenomenon in Nature" explores the intriguing realm of natural wave phenomena and their practical implications within the natural environment. The text commences by providing an introductory overview of waves, wherein waves are defined as the transmission of energy through a medium without the physical displacement of matter. Waves are widely observed in the natural environment, exerting influence on diverse facets of existence. The chapter examines three primary classifications of waves that are observed in the natural world. Mechanical waves, such as water waves and seismic waves, necessitate a medium through which they can propagate. Electromagnetic waves, including light waves and radio waves, have the ability to propagate through a vacuum medium. Matter waves, which are linked to quantum phenomena, serve as a demonstration of the dualistic nature exhibited by particles. This study focuses on the analysis of water waves, encompassing both oceanic waves and ripples found in lakes and ponds. This paper examines the processes involved in the generation and transmission of ocean waves, along with an analysis of the devastating impact that tsunamis can have. This study examines the phenomena of surface tension and capillary waves, with a particular focus on their interactions with wind and gravity in smaller aquatic environments. The chapter provides an in-depth exploration of seismic waves, presenting the field of seismology and discussing the structure and composition of the Earth. The article examines different categories of seismic waves, including P-waves and S-waves, and their characteristics and responses in the context of seismic events. The significance of seismic waves on structures is underscored, highlighting the criticality of earthquake preparedness and engineering interventions. This study investigates atmospheric waves, with a specific emphasis on sound waves within the atmosphere and atmospheric gravity waves. This study investigates the characteristics of sound transmission in the air and its interaction with atmospheric phenomena, such as inversions of temperature and wind gradients. The significance of atmospheric gravity waves in shaping weather patterns and climate variability is emphasized. This study examines the significance of electromagnetic waves in natural phenomena, with a specific focus on sunlight and solar waves. This chapter examines the solar spectrum, its interaction with the Earth's atmosphere, and the occurrence of solar flares and their effects on space weather and technological infrastructure. The study of biological and biological-physical waves encompasses an examination of phenomena such as bioluminescence and the transmission of light signals among organisms. This essay elucidates the importance of bioluminescence in various biological contexts, including communication, camouflage, and mate attraction. This study investigates sound-based communication within the animal kingdom, specifically focusing on vocalizations observed in avian, mammalian, and insect species. This study investigates the distinctive capacity of select animal species, such as bats and marine mammals, to employ echolocation as a means of navigation and prey detection. The chapter emphasizes the multifaceted and indispensable function of waves in the natural realm. Waves possess the ability to not only influence physical phenomena but also hold significant importance in the realms of communication, navigation, and the survival mechanisms of diverse biological systems. The investigation of these wave phenomena contributes to the advancement of our comprehension regarding the complex interrelationships within the natural realm.

REVIEW QUESTIONS

1. What are the three types of waves discussed in nature, and how do they differ from each other?
2. Explain the characteristics of mechanical waves and provide examples of where they can be found in nature.
3. Discuss the properties of electromagnetic waves and their significance in natural phenomena.
4. How are matter waves different from mechanical and electromagnetic waves? Provide examples of matter waves.
5. Describe the various types of water waves, with a focus on ocean waves and ripples in lakes and ponds.
6. What are seismic waves, and how are they related to earthquakes?

Explain the different types of seismic waves.
7. Discuss the propagation of sound waves in the atmosphere and their relevance to atmospheric phenomena.
8. Explain the concept of atmospheric gravity waves and their role in atmospheric dynamics.
9. How do natural sources, such as sunlight and celestial bodies, produce electromagnetic waves in nature?
10. Explore the biological and biological-physical waves, particularly bioluminescence, light communication, and sound communication in the animal kingdom.

REFERENCES

1. Afraimovich, E. L. (1981). Cepstral analysis of broad-band radio emission-New possibilities in radio astronomy. *Astronomy and Astrophysics, vol. 97, no. 2, Apr. 1981, p. 366–372.*, *97*, 366–372.
2. Ambroselli, M., Poulos, P., & Roychoudhuri, C. (2011). Nature of EM waves as observed and reported by detectors for radio, visible, and gamma frequencies. *The Nature of Light: What are Photons? IV*, *8121*, 466–474.
3. Aspler, L. B., Chiarenzelli, J. R., & Bursey, T. L. (1994). Ripple marks in quartz arenites of the Hurwitz Group, Northwest Territories, Canada; evidence for sedimentation in a vast, early Proterozoic, shallow, fresh-water lake. *Journal of Sedimentary Research*, *64*(2), 282–298.
4. Bai, J., Zong, X., Ma, Y., Wang, B., Zhao, C., Yang, Y., ... & Song, T. (2022). Long-Term Variations in Global Solar Radiation and Its Interaction with Atmospheric Substances at Qomolangma. *International Journal of Environmental Research and Public Health*, *19*(15), 8906.
5. Bakhmet'eva, N. V., Grigoriev, G. I., Tolmacheva, A. V., & Kalinina, E. E. (2018). Atmospheric turbulence and internal gravity waves examined by the method of artificial periodic irregularities. *Russian Journal of Physical Chemistry B*, *12*, 510–521.
6. Billy, J., Josse, V., Zuo, Z., Bernard, A., Hambrecht, B., Lugan, P., ... & Aspect, A. (2008). Direct observation of Anderson localization of matter waves in a controlled disorder. *Nature*, *453*(7197), 891–894.
7. Brinkløv, S. M., Jakobsen, L., & Miller, L. A. (2022). Echolocation in bats, odontocetes, birds, and insectivores. *Exploring Animal Behavior Through Sound: Volume*, 419, 97–99.
8. Burchardt, L. S., & Knörnschild, M. (2020). Comparison of methods for rhythm analysis of complex animals' acoustic signals. *PLOS Computational Biology*, *16*(4), e1007755.
9. Burke, A., & Crocker, A. (2020). "Making" Waves: How Young Learners Connect to Their Natural World through Third Space. *Education Sciences*, *10*(8), 203.
10. Chavez, F. P., Strutton, P. G., & McPhaden, M. J. (1998). Biological-physical coupling in the central equatorial Pacific during the onset of the 1997–98 El Niño. *Geophysical Research Letters*, *25*(19), 3543–3546.

11. Cusick, K. D., & Widder, E. A. (2020). Bioluminescence and toxicity as driving factors in harmful algal blooms: ecological functions and genetic variability. *Harmful Algae*, *98*, 101850.

12. Dagkesamanskii, R. D., & Zheleznykh, I. M. (1989). A radio astronomy method of detecting neutrinos and other superhigh-energy elementary particles. *Pisma v Zhurnal Eksperimentalnoi i Teoreticheskoi Fiziki*, *50*, 233–235.

13. Deryck, L. (1978). Natural propagation of electromagnetic waves in tunnels. *IEEE Transactions on Vehicular Technology*, *27*(3), 145–150.

14. Dong, S., Tang, W., Hu, P., Zhao, X., Zhang, X., Han, J., & Hu, P. (2019). Achieving excellent electromagnetic wave absorption capabilities by construction of MnO nanorods on porous carbon composites derived from natural wood via a simple route. *ACS Sustainable Chemistry & Engineering*, *7*(13), 11795–11805.

15. Dörfler, W., Hochbruck, M., Köhler, J., Rieder, A., Schnaubelt, R., & Wieners, C. (2023). *Wave Phenomena: Mathematical Analysis and Numerical Approximation.* Springer Nature. (Vol. 49, pp. 87–90).

16. Eisenmenger, W. (1959). Dynamic properties of the surface tension of water and aqueous solutions of surface active agents with standing capillary waves in the frequency range from 10 kc/s to 1.5 Mc/s. *Acta Acustica United with Acustica*, *9*(4), 327–340.

17. Evans, W. E., & Awbrey, F. T. (1988). Natural history aspects of marine mammal echolocation: feeding strategies and habitat. In *Animal Sonar: Processes and Performance.* Boston, MA: Springer US. (Vol. 2, pp. 521–534).

18. Fratamico, A., Tocquin, P., & Franck, F. (2016). The chlorophyll a fluorescence induction curve in the green microalga Haematococcus pluvialis: further insight into the nature of the P–S–M fluctuation and its relationship with the "low-wave" phenomenon at steady-state. *Photosynthesis Research*, *128*, 271–285.

19. Frölicher, T. L., & Laufkötter, C. (2018). Emerging risks from marine heat waves. *Nature Communications*, *9*(1), 650.

20. Gattuso, J. P., Gentili, B., Duarte, C. M., Kleypas, J. A., Middelburg, J. J., & Antoine, D. (2006). Light availability in the coastal ocean: impact on the distribution of benthic photosynthetic organisms and their contribution to primary production. *Biogeosciences*, *3*(4), 489–513.

21. Ginsberg, N. S., Garner, S. R., & Hau, L. V. (2007). Coherent control

of optical information with matter wave dynamics. *Nature*, *445*(7128), 623–626.

22. Greiner, M., Mandel, O., Hänsch, T. W., & Bloch, I. (2002). Collapse and revival of the matter wave field of a Bose–Einstein condensate. *Nature*, *419*(6902), 51–54.
23. Gruszecki, P., Lyubchanskii, I. L., Guslienko, K. Y., & Krawczyk, M. (2021). Local non-linear excitation of sub-100 nm bulk-type spin waves by edge-localized spin waves in magnetic films. *Applied Physics Letters*, *118*(6), 587–590.
24. Hanks, T. C. (1977). Earthquake stress drops, ambient tectonic stresses and stresses that drive plate motions. *Stress in the Earth*, 441–458.
25. Hocke, K., & Schlegel, K. (1996, September). A review of atmospheric gravity waves and travelling ionospheric disturbances: 1982–1995. In *Annales geophysicae* (Vol. 14, pp. 917–920).
26. Iversen, M. M., Zhu, H., Zhou, W., Della Santina, C. C., Carey, J. P., & Rabbitt, R. D. (2018). Sound abnormally stimulates the vestibular system in canal dehiscence syndrome by generating pathological fluid-mechanical waves. *SCIENTIFIC REPOrts*, *8*(1), 10257.
27. Jeffreys, H. (1925). On the formation of water waves by wind. *Proceedings of the royal society of London. Series A, Containing Papers of a Mathematical and Physical Character*, *107*(742), 189–206.
28. Jess, D. B., Shelyag, S., Mathioudakis, M., Keys, P. H., Christian, D. J., & Keenan, F. P. (2012). Propagating wave phenomena detected in observations and simulations of the lower solar atmosphere. *The Astrophysical Journal*, *746*(2), 183.
29. Kino, N., & Ueno, T. (2008). Evaluation of acoustical and non-acoustical properties of sound absorbing materials made of polyester fibres of various cross-sectional shapes. *Applied Acoustics*, *69*(7), 575–582.
30. Kuwayama, Y., Hirose, K., Sata, N., & Ohishi, Y. (2008). Phase relations of iron and iron–nickel alloys up to 300 GPa: Implications for composition and structure of the Earth's inner core. *Earth and Planetary Science Letters*, *273*(3–4), 379–385.
31. Kuznetsov, S. V. (2011). Seismic waves and seismic barriers. *Acoustical Physics*, *57*, 420–426.
32. Latief, H., & Hadi, S. (2007). The role of forests and trees in protecting coastal areas against tsunamis. In *Regional Technical Workshop on*

Coastal Protection in the Aftermath of the Indian Ocean Tsunami What Role for Forests Trees (Vol. 2, pp. 5–32).

33. Liang, C., Gou, Y., Wu, L., Zhou, J., Kang, Z., Shen, B., & Wang, Z. (2016). Nature of electromagnetic-transparent SiO2 shell in hybrid nanostructure enhancing electromagnetic attenuation. *The Journal of Physical Chemistry C, 120*(24), 12967–12973.
34. Luco, J. E., & Wong, H. L. (1982). Response of structures to nonvertically incident seismic waves. *Bulletin of the Seismological Society of America*, *72*(1), 275–302.
35. Luo, J., Dai, Z., Feng, M., Chen, X., Sun, C., & Xu, Y. (2022). Hierarchically porous carbon derived from natural Porphyra for excellent electromagnetic wave absorption. *Journal of Materials Science & Technology*, *129*, 206–214.
36. MacDowell, L. G., Benet, J., Katcho, N. A., & Palanco, J. M. (2014). Disjoining pressure and the film-height-dependent surface tension of thin liquid films: New insight from capillary wave fluctuations. *Advances in Colloid and Interface Science*, *206*, 150–171.
37. Mach, L., Ponting, J., Brown, J., & Savage, J. (2020). Riding waves of intra-seasonal demand in surf tourism: Analysing the nexus of seasonality and 21st century surf forecasting technology. *Annals of Leisure Research*, *23*(2), 184–202.
38. Macquart, J. P., Prochaska, J. X., McQuinn, M., Bannister, K. W., Bhandari, S., Day, C. K., ... & Tejos, N. (2020). A census of baryons in the Universe from localized fast radio bursts. *Nature*, *581*(7809), 391–395.
39. Malladi, V. V. N. S., Avirovik, D., Priya, S., & Tarazaga, P. (2015). Characterization and representation of mechanical waves generated in piezo-electric augmented beams. *Smart Materials and Structures*, *24*(10), 105026.
40. Mani, R. G., Smet, J. H., von Klitzing, K., Narayanamurti, V., Johnson, W. B., & Umansky, V. (2002). Zero-resistance states induced by electromagnetic-wave excitation in GaAs/AlGaAs heterostructures. *Nature*, *420*(6916), 646–650.
41. Marler, P. (1967). Animal Communication Signals: We are beginning to understand how the structure of animal signals relates to the function they serve. *Science*, *157*(3790), 769–774.
42. Masten, A. S. (2007). Resilience in developing systems: Progress and

promise as the fourth wave rises. *Development and psychopathology*, *19*(3), 921–930.

43. McElroy, K., Simmonds, R. W., Hoffman, J. E., Lee, D. H., Orenstein, J., Eisaki, H., ... & Davis, J. C. (2003). Relating atomic-scale electronic phenomena to wave-like quasiparticle states in superconducting Bi2Sr2CaCu2O8+ δ. *Nature*, *422*(6932), 592–596.
44. Medvedev, A. S., & Klaassen, G. P. (2000). Parameterization of gravity wave momentum deposition based on nonlinear wave interactions: Basic formulation and sensitivity tests. *Journal of Atmospheric and Solar-Terrestrial Physics*, *62*(11), 1015–1033.
45. Molnar, P. (1979). Earthquake recurrence intervals and plate tectonics. *Bulletin of the Seismological Society of America*, *69*(1), 115–133.
46. Mozer, F. S., Bonnell, J. W., Hanson, E. L. M., Gasque, L. C., & Vasko, I. Y. (2021). Nonlinear ion-acoustic waves, ion holes, and electron holes in the near-Sun solar wind. *The Astrophysical Journal*, *911*(2), 89.
47. Mukhlas, N. A., Mohd Zaki, N. I., Abu Husain, M. K., Syed Ahmad, S. Z. A., & Najafian, G. (2022). Numerical formulation based on ocean wave mechanics for offshore structure analysis–a review. *Ships and Offshore Structures*, *17*(8), 1731–1742.
48. Müller, H., Peters, A., & Chu, S. (2010). A precision measurement of the gravitational redshift by the interference of matter waves. *Nature*, *463*(7283), 926–929.
49. Nestola, F., Nimis, P., Ziberna, L., Longo, M., Marzoli, A., Harris, J. W., ... & Fedortchouk, Y. (2011). First crystal-structure determination of olivine in diamond: Composition and implications for provenance in the Earth's mantle. *Earth and Planetary Science Letters*, *305*(1–2), 249–255.
50. Onorato, M., Waseda, T., Toffoli, A., Cavaleri, L., Gramstad, O., Janssen, P. A. E. M., ... & Trulsen, K. (2009). Statistical properties of directional ocean waves: the role of the modulational instability in the formation of extreme events. *Physical Review Letters*, *102*(11), 114502.
51. Paraskevis, D., Kostaki, E. G., Alygizakis, N., Thomaidis, N. S., Cartalis, C., Tsiodras, S., & Dimopoulos, M. A. (2021). A review of the impact of weather and climate variables to COVID-19: In the absence of public health measures high temperatures cannot probably mitigate

outbreaks. *Science of the Total Environment*, *768*, 144578.

52. Phillips, O. M. (1985). Spectral and statistical properties of the equilibrium range in wind-generated gravity waves. *Journal of Fluid Mechanics*, *156*, 505–531.
53. Podolefsky, N. S., & Finkelstein, N. D. (2007). Analogical scaffolding and the learning of abstract ideas in physics: An example from electromagnetic waves. *Physical Review Special Topics-Physics Education Research*, *3*(1), 010109.
54. Polet, J., & Kanamori, H. (2000). Shallow subduction zone earthquakes and their tsunamigenic potential. *Geophysical Journal International*, *142*(3), 684–702.
55. Poole, J. A., Barnes, C. S., Demain, J. G., Bernstein, J. A., Padukudru, M. A., Sheehan, W. J., ... & Nel, A. E. (2019). Impact of weather and climate change with indoor and outdoor air quality in asthma: A Work Group Report of the AAAAI Environmental Exposure and Respiratory Health Committee. *Journal of Allergy and Clinical Immunology*, *143*(5), 1702–1710.
56. Rahmani, Y., Alizadeh, M. M., Schuh, H., Wickert, J., & Tsai, L. C. (2020). Probing vertical coupling effects of thunderstorms on lower ionosphere using GNSS data. *Advances in Space Research*, *66*(8), 1967–1976.
57. Rossby, C. G. (1945). On the propagation of frequencies and energy in certain types of oceanic and atmospheric waves. *Journal of the Atmospheric Sciences*, *2*(4), 187–204.
58. Scarpa, F., Chowdhury, R., Kam, K., Adhikari, S., & Ruzzene, M. (2011). Dynamics of mechanical waves in periodic graphene nanoribbon assemblies. *Nanoscale Research Letters*, *6*, 1–10.
59. Schöbel, S., Schmitt, A., Benner, D., Saqr, M., Janson, A., & Leimeister, J. M. (2023). Charting the Evolution and Future of Conversational Agents: A Research Agenda Along Five Waves and New Frontiers. *Information Systems Frontiers*, 10(5), 1–26.
60. Schumm, T., Hofferberth, S., Andersson, L. M., Wildermuth, S., Groth, S., Bar-Joseph, I., ... & Krüger, P. (2005). Matter-wave interferometry in a double well on an atom chip. *Nature Physics*, *1*(1), 57–62.
61. Semblat, J. F., Kham, M., & Bard, P. Y. (2008). Seismic-wave propagation in alluvial basins and influence of site-city interaction. *Bulletin of the Seismological Society of America*, *98*(6), 2665–2678.

62. Sleep, N. H., & Nakata, N. (2015). Nonlinear attenuation from the interaction between different types of seismic waves and interaction of seismic waves with shallow ambient tectonic stress. *Geochemistry, Geophysics, Geosystems*, *16*(7), 2336–2363.
63. Smith, M. D. (2011). An ecological perspective on extreme climatic events: a synthetic definition and framework to guide future research. *Journal of Ecology*, *99*(3), 656–663.
64. Sonak, S., Pangam, P., & Giriyan, A. (2008). Green reconstruction of the tsunami-affected areas in India using the integrated coastal zone management concept. *Journal of Environmental Management*, *89*(1), 14–23.
65. Spivak, A. A., Rybnov, Y. S., Soloviev, S. P., & Kharlamov, V. A. (2018). Acoustic and electric precursors of strong thunderstorm events under megalopolis conditions. *Izvestiya, Atmospheric and Oceanic Physics*, *54*, 738–744.
66. Stopa, J. E., Sutherland, P., & Ardhuin, F. (2018). Strong and highly variable push of ocean waves on Southern Ocean sea ice. *Proceedings of the National Academy of Sciences*, *115*(23), 5861–5865.
67. Sumi, N., & Ashida, F. (2003). Solution for thermal and mechanical waves in a piezoelectric plate by the method of characteristics. *Journal of thermal stresses*, *26*(11–12), 1113–1123.
68. Szczuciński, W., Chaimanee, N., Niedzielski, P., Rachlewicz, G., Saisuttichai, D., Tepsuwan, T., ... & Siepak, J. (2006). Environmental and Geological Impacts of the 26 December 2004 Tsunami in Coastal Zone of Thailand--Overview of Short and Long-Term Effects. *Polish Journal of Environmental Studies*, *15*(5), 247–250.
69. Taflove, A., & Umashankar, K. R. (1989). Review of FD-TD numerical modeling of electromagnetic wave scattering and radar cross section. *Proceedings of the IEEE*, *77*(5), 682–699.
70. Tongchai, A., Sharma, M. D., Johnston, I. D., Arayathanitkul, K., & Soankwan, C. (2011). Consistency of students' conceptions of wave propagation: Findings from a conceptual survey in mechanical waves. *Physical Review Special Topics-Physics Education Research*, *7*(2), 020101.
71. Tongchai, A., Sharma, M. D., Johnston, I. D., Arayathanitkul, K., & Soankwan, C. (2009). Developing, evaluating and demonstrating the use of a conceptual survey in mechanical waves. *International Journal*

of Science Education, *31*(18), 2437–2457.

72. Tromans, P. S., Anaturk, A. R., & Hagemeijer, P. (1991, August). A new model for the kinematics of large ocean waves-application as a design wave. In *ISOPE International Ocean and Polar Engineering Conference*. ISOPE. (Vol. 2, pp. 282–285).
73. Tsai, Y. Y., Tsai, J. Y., & I, L. (2016). Generation of acoustic rogue waves in dusty plasmas through three-dimensional particle focusing by distorted waveforms. *Nature Physics*, *12*(6), 573–577.
74. Wang, X., Luo, X., & Huang, Z. (2020). Hybrid metamaterials enable multifunctional manipulation of mechanical waves on solid-fluid interfaces. *Applied Physics Letters*, *117*(6), 5872.
75. Werner, A. D., Bakker, M., Post, V. E., Vandenbohede, A., Lu, C., Ataie-Ashtiani, B., ... & Barry, D. A. (2013). Seawater intrusion processes, investigation and management: Recent advances and future challenges. *Advances in water resources*, *51*, 3–26.
76. Wiley, R. H., & Richards, D. G. (1978). Physical constraints on acoustic communication in the atmosphere: implications for the evolution of animal vocalizations. *Behavioral Ecology and Sociobiology*, *3*, 69–94.
77. Yiğit, E., Knížová, P. K., Georgieva, K., & Ward, W. (2016). A review of vertical coupling in the Atmosphere–Ionosphere system: Effects of waves, sudden stratospheric warmings, space weather, and of solar activity. *Journal of Atmospheric and Solar-Terrestrial Physics*, *141*, 1–12.
78. Zaikin, A. N., & Zhabotinsky, A. M. (1970). Concentration wave propagation in two-dimensional liquid-phase self-oscillating system. *Nature*, *225*(5232), 535–537.
79. Zhao, J., Wang, T., Graham, D. B., He, J., Liu, W., Dunlop, M. W., & Wu, D. (2020). Identification of the Nature of Electromagnetic Waves near the Proton-cyclotron Frequency in Solar-terrestrial Plasmas. *The Astrophysical Journal*, *890*(1), 17.

CHAPTER 3

Sound Waves

Contents

3.1. INTRODUCTION

Sound is an integral part of our everyday lives, surrounding us with a rich tapestry of auditory experiences. From the soothing melodies of music to the bustling noise of a city, sound waves play a fundamental role in our perception of the world. In this comprehensive exploration of sound waves, we will delve into their definition, characteristics, and the significance they hold in various applications. Sound waves are mechanical waves that travel through a medium, such as air, water, or solids, as a result of vibrations or oscillations of particles in the medium. When an object vibrates, it causes the adjacent particles of the medium to also vibrate, creating a chain reaction that propagates the sound energy through the medium. This transmission of energy as a wave of pressure variations results in the formation of sound waves (Tomonaga, 1950).

Sound waves can be described in terms of their frequency, wavelength, and amplitude. The frequency of a sound wave refers to the number of oscillations or cycles it completes per second and is measured in Hertz (Hz). The wavelength is the distance between two consecutive points of the wave that are in phase, while the amplitude represents the maximum displacement of the particles from their equilibrium position (Wittmann et al., 2003).

Following are the characteristics of sound waves:

- ***Speed*:** The speed of sound waves depends on the properties of the medium through which they travel. In dry air at room temperature, sound travels at approximately 343 meters per second (m/s), but this speed can vary in different materials. For instance, sound travels faster in water and even faster in solids like steel (Kasper et al., 2015).
- ***Propagation*:** Sound waves propagate in all directions from their source in a spherical pattern. As they move away from the source, the energy of the sound waves spreads out, causing the intensity of the sound to decrease with distance (Kaji & Okazaki, 1970).
- ***Reflection and Refraction*:** Sound waves can undergo reflection and refraction when they encounter obstacles or pass through different mediums. Reflection occurs when sound waves bounce off a surface, while refraction involves a change in direction as sound waves cross from one medium to another with varying properties (Sitdikova & Gimaltdinov, 2022).
- ***Interference*:** When two or more sound waves meet, they can

interfere constructively, resulting in an increase in amplitude, or destructively, causing a decrease in amplitude. This phenomenon of sound wave interference plays a crucial role in various audio technologies and music (Daigle, 1979).

- ***Resonance*:** Resonance is a phenomenon in which an object vibrates with a greater amplitude when exposed to a sound wave that matches its natural frequency. This property is harnessed in musical instruments to amplify and produce distinct sounds (Li et al., 2022).
- ***Frequency Spectrum*:** The range of audible frequencies for humans is typically between 20 Hz to 20,000 Hz. This range constitutes the audible frequency spectrum and serves as the basis for various applications, including music and communication (Lin et al., 2022).

The significance of sound waves extends to numerous domains, with applications that are both practical and essential to our way of life:

- ***Communication*:** Sound waves play a vital role in communication systems. From speech and telephone calls to radio broadcasts and public announcements, sound carries information over vast distances, enabling effective communication among individuals and societies (Awad & Barak, 2018).
- ***Music and Entertainment*:** Sound waves are at the heart of music and the entertainment industry. Musical instruments produce sound through the vibration of strings, air columns, or membranes, while speakers and headphones recreate sound for our listening pleasure (Akins, 2006).
- ***Medicine*:** Sound waves find extensive use in medical imaging techniques like ultrasound. Ultrasound waves, beyond our hearing range, can penetrate the body and create images used in diagnosing medical conditions and monitoring fetal development during pregnancy (Katz et al., 2020).
- ***Sonar Technology*:** Sonar systems use sound waves to detect and locate objects underwater. This technology is employed in navigation, oceanography, and even fish detection in the fishing industry (Santos & Macedo Filho, 2006).
- ***Acoustics and Architecture*:** Sound waves influence architectural design to create spaces with optimal acoustics. Concert halls, theaters, and recording studios are carefully engineered to ensure

excellent sound quality and minimize undesirable echoes (Elson, 1921).

- ***Noise Control*:** Understanding sound waves is crucial for noise control in various environments. Engineers employ sound-absorbing materials and design techniques to minimize noise pollution in residential, industrial, and urban settings (Sawada & Ohsumi, 2000).
- ***Security and Surveillance*:** Acoustic sensors and devices are used in security systems for detecting sounds associated with intrusions or suspicious activities. These systems enhance surveillance and aid in safeguarding sensitive areas (Saha et al., 2017).
- ***Earthquake Detection*:** Seismometers use sound waves (earthquake waves) to detect and measure seismic activity, providing critical information for earthquake monitoring and early warning systems (Heki et al., 2006).
- ***Non-Destructive Testing (NDT)*:** In industries such as aerospace and manufacturing, sound waves are used for NDT to inspect materials and structures without causing damage. Ultrasonic testing is one such technique that assesses flaws or irregularities in materials (Jolly et al., 2015).
- ***Sonomaterials and Acoustic Levitation*:** Researchers explore the use of sound waves in manipulating and levitating small objects, leading to advancements in materials science and technology.

Sound waves are a fascinating and versatile aspect of physics, impacting our lives in countless ways. Their propagation through various media, characteristics, and applications in communication, entertainment, medicine, industry, and other fields make them an indispensable element of our modern world. Understanding sound waves not only enhances our appreciation of the auditory experiences we encounter daily but also drives innovation and development across multiple industries (Andrade et al., 2020).

3.2. SPEED OF SOUND

Sound, as a mechanical wave, travels through various media at a specific rate known as the speed of sound. This fundamental property of sound waves is essential in understanding how sound propagates and interacts with its environment. In this section, we will delve into the definition, concept, factors influencing the speed of sound, methods of calculation and measurement, and real-world significance of this crucial aspect of sound waves. The speed of sound refers to the rate at which sound waves travel through a specific medium. In a medium with particles, such as air, water, or solids, sound waves propagate through the transfer of energy from one particle to the next. As an object vibrates, it causes neighboring particles to oscillate, creating areas of compression and rarefaction. These alternating regions of high and low pressure from the sound wave, which moves outward from the source in all directions (Figure 3.1).

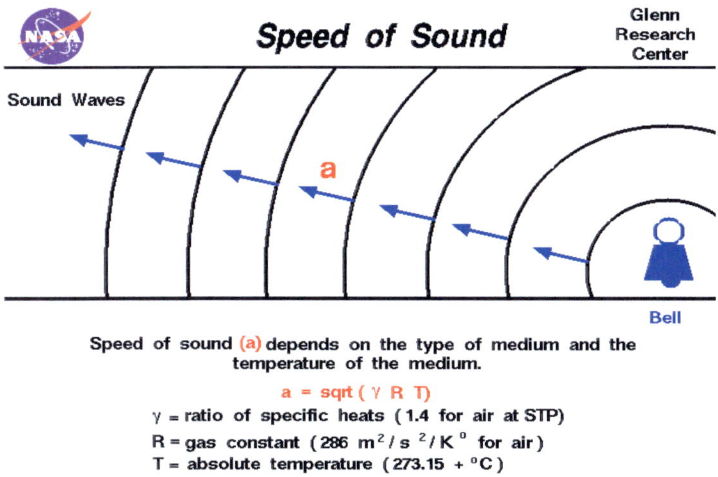

Figure 3.1. Illustration of the speed of sound (Source: NASA, Creative Commons License).

The speed of sound depends on the physical properties of the medium through which it travels, such as density, elasticity, and temperature. Generally, sound travels faster in denser and more elastic materials. For example, sound waves propagate faster in water than in air due to the higher density of water molecules and the stronger intermolecular forces between them.

The concept of the speed of sound is a fundamental aspect of wave theory and is governed by the relationship between wavelength (λ), frequency (f), and speed (v):

$$v = f * \lambda$$

where: v = Speed of sound f = Frequency of the sound wave λ = Wavelength of the sound wave

The speed of sound in a specific medium is a constant value at a given temperature and pressure. However, it can change when the medium's properties, such as temperature and humidity, undergo alterations (Greenspan & Tschiegg, 1957).

3.2.1. Factors Affecting the Speed of Sound

Following are the factors affecting the speed of sound:

- ***Temperature*:** Temperature is the most significant factor affecting the speed of sound in a medium. In general, sound travels faster in warmer temperatures and slower in colder temperatures. As the temperature increases, the molecules in the medium gain more kinetic energy, leading to faster vibrations and greater speed of sound (Linder, 1993).
- ***Density of the Medium*:** The density of the medium also influences the speed of sound. In denser materials, the particles are closer together, allowing sound waves to propagate more quickly. For instance, sound travels faster in liquids and solids, which have higher densities compared to gases.
- ***Elasticity of the Medium*:** Elasticity refers to the ability of a material to deform and return to its original shape when a force is applied. A more elastic medium allows sound waves to travel faster since the particles can quickly bounce back after compression or rarefaction (Blake, 1848).

Tip

The speed of sound in a medium depends on factors such as temperature, density, and elasticity. Higher temperatures and greater elasticity generally lead to faster sound propagation.

- ***Pressure*:** While pressure can influence the speed of sound, its effect is generally negligible under normal conditions. In gases, a higher pressure increases the density of the medium, leading to a slight increase in the speed of sound.
- ***Humidity*:** Humidity, or the amount of moisture present in the air, can have a minor impact on the speed of sound in gases. Moist air is less dense than dry air, and this slight decrease in density can affect the speed of sound slightly.
- ***Composition of the Medium*:** The atomic or molecular composition of the medium can also impact the speed of sound.

For example, sound travels differently in a mixture of gases than in a pure gas due to the differing masses and intermolecular forces (Njeh et al., 1999).

3.2.2. Calculations and Measurement of Speed of Sound

The speed of sound in a specific medium can be calculated using the relationship mentioned earlier:

$v = f * \lambda$

- To measure the speed of sound experimentally, various methods can be employed:
- ***Echo Method*:** In this method, a loud sound is produced at a known distance from a reflecting surface, such as a wall. The time taken for the sound to travel to the surface and return as an echo is measured. Dividing the total distance traveled by the time taken gives the speed of sound (Thol et al., 2017).
- ***Resonance Tube Method*:** A resonating tube is used to produce standing waves of sound. By changing the length of the tube and finding the position where resonance occurs, the wavelength of the sound wave can be determined. Combining this with the frequency of the sound wave, the speed of sound can be calculated.
- ***Time-of-Flight Method*:** In this approach, a sound pulse is sent and detected at known distances. By measuring the time taken for the sound to travel to the detector, the speed of sound can be calculated.
- ***Interference Method*:** In this technique, two sources of sound emit waves of slightly different frequencies. The interference pattern of these waves is measured, and from the pattern, the wavelength of the sound is deduced. With the frequency known, the speed of sound can be determined (Aksenov & Kuznetsov, 2021).

3.2.3. Examples and Real-World Significance

The speed of sound has significant real-world implications and finds applications in various fields:

- ***Weather Forecasting*:** The speed of sound can be used to calculate the distance of a lightning strike based on the time interval between seeing the lightning and hearing the thunder. This information aids in estimating the proximity of a storm and predicting its potential impact (Cressman, 1965).

- ***Aviation and Aerospace*:** Knowledge of the speed of sound is crucial in aviation and aerospace engineering. Aircraft must consider the speed of sound when traveling at high speeds, as breaking the sound barrier can cause shock waves and aerodynamic challenges.
- ***Sonar Technology*:** In naval and oceanographic applications, sonar systems use the speed of sound in water to calculate distances and detect underwater objects, such as submarines or underwater geological features.
- ***Medical Imaging*:** In medical ultrasound imaging, the speed of sound in soft tissues is used to determine the depth of structures within the body. This aids in diagnosing medical conditions and monitoring fetal development during pregnancy (Stähli et al., 2020).
- ***Geology and Seismology*:** The speed of sound in various geological materials provides valuable information for understanding the Earth's subsurface and detecting seismic activity. Seismologists use this data to analyze earthquakes and investigate the Earth's structure.
- ***Communication Technology*:** Understanding the speed of sound is essential in the development of communication systems, such as telephony and radio, where delays in signal transmission can affect overall performance.
- ***Material Testing and Quality Control*:** The speed of sound is used in non-destructive testing (NDT) to assess the integrity of materials in industries like aerospace and manufacturing. Ultrasonic testing, for example, determines flaws or irregularities without damaging the material (Maia et al., 2021).
- ***Architecture and Acoustics*:** In architectural design, the speed of sound plays a role in determining the optimal shape and dimensions of spaces, ensuring the desired acoustic properties of concert halls, theaters, and recording studios.
- ***Underwater Communication*:** In underwater environments, the speed of sound in water is utilized for communication between submarines and other underwater vehicles.

The speed of sound is a fundamental property of sound waves that influences their propagation and interaction with different media. Temperature, density, elasticity, and composition of the medium are among the factors affecting the speed of sound. By understanding the speed of sound, we can apply this knowledge to various fields, including aviation, medical imaging, geology, communication technology, and more. The practical significance of the speed of sound highlights its critical role in shaping modern technology and enhancing our understanding of the natural world (Stojanovic & Preisig, 2009).

3.3. SOUND INTENSITY

Sound intensity is a fundamental characteristic of sound waves that quantifies the amount of sound energy transmitted per unit area in a specific direction. It plays a crucial role in understanding the strength, loudness, and potential effects of sound in various applications. In this section, we will explore the concept of sound intensity, the decibel scale used to measure it, the logarithmic nature of intensity, methods of measurement and units, and the diverse applications and effects of sound intensity. Sound intensity refers to the amount of acoustic energy passing through a given area per unit of time. It is typically measured in watts per square meter (W/m^2) and quantifies how "loud" a sound is at a particular point in space. As sound waves propagate outward from the source, the intensity of the sound decreases with increasing distance from the source due to energy spreading out over a larger area (Figure 3.2).

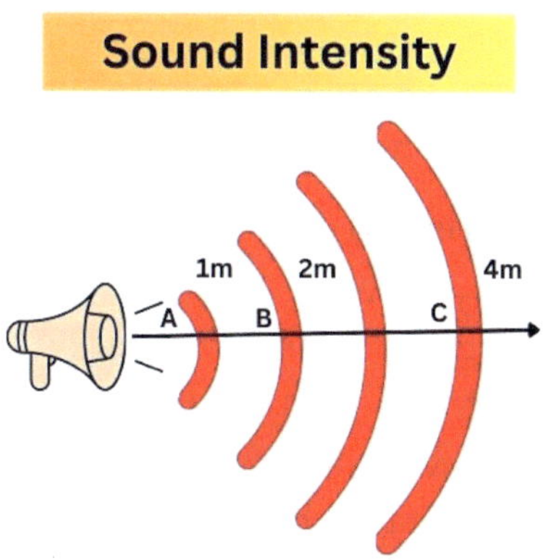

Figure 3.2. Sound Intensity is referred to the sound per unit area perpendicular to the path of the sound waves (Source: Muskan Shafi, Creative Commons Licnese).

The human perception of sound intensity is closely related to the perceived loudness (Figure 3.3). Higher sound intensities are perceived as louder, while lower intensities are perceived as softer. However, the relationship between intensity and perceived loudness is not linear but follows a logarithmic scale, which is the basis for the decibel (dB) scale used to measure sound intensity (Jacobsen & de Bree, 2005).

3.3.1. Decibel Scale and Logarithmic Nature of Sound Intensity

The decibel (dB) scale is a logarithmic scale used to express sound intensity levels. The decibel scale allows us to compress a wide range of intensities into a more manageable scale, making it easier to represent the vast range of sounds humans can hear. The formula for calculating sound intensity level in decibels is as follows:

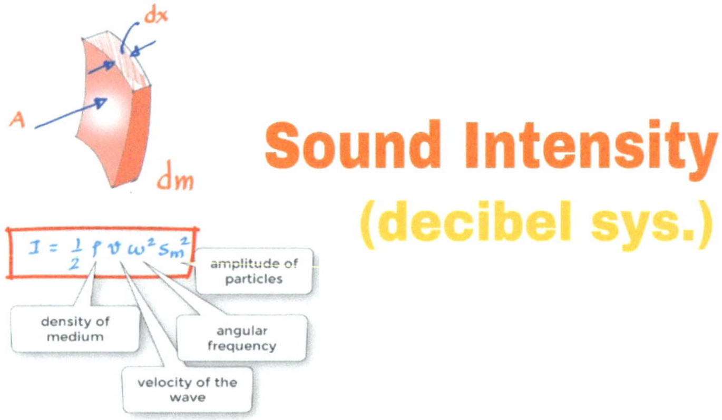

Figure 3.3. Sound Intensity and decibel scale (the logarithmic scale) (Source: The Science Cube, Creative Commons License).

Sound Intensity Level (in dB) = 10 * log10(I/I_0)

where: I = Sound Intensity in W/m² I_0 = Reference Sound Intensity (the threshold of human hearing) = 1 x 10^(-12) W/m²

On the decibel scale, the threshold of human hearing is assigned 0 dB, and the intensity level of any sound is expressed relative to this threshold. Sounds above the threshold have positive dB values, while sounds below the threshold have negative dB values. It's important to note that a 10 dB increase represents a sound that is perceived as being approximately twice as loud, and a 20 dB increase represents a tenfold increase in perceived loudness (Robin & Plante, 2022).

3.3.2. Measurement and Units of Sound Intensity

Measuring sound intensity is a critical aspect of various fields, such as environmental monitoring, industrial noise control, and audiology. Sound intensity meters, also known as sound level meters, are devices used to measure sound intensity levels in decibels.

Sound intensity meters consist of a microphone that detects sound waves and converts them into electrical signals. These signals are then processed and displayed as sound intensity levels in decibels on the meter's screen. The meter may have different settings for different frequency weightings, such as A-weighting, which approximates the sensitivity of the human ear to different frequencies.

The units of sound intensity, as mentioned earlier, are in watts per square meter (W/m^2). However, in practice, sound intensity levels are expressed in decibels (dB) to simplify representation and perception (Crocker & Arenas, 2003).

3.3.3. Applications and Effects of Sound Intensity

The concept of sound intensity and its measurement have numerous applications in various fields, including:

- ***Environmental Noise Monitoring*:** Sound intensity measurements are essential for monitoring and assessing environmental noise levels, especially in urban areas, construction sites, and industrial zones. This information helps in implementing noise control measures and ensuring compliance with noise regulations (Kuwano et al., 1989).
- ***Occupational Health and Safety*:** In industrial settings, sound intensity measurements are used to evaluate workplace noise exposure. This helps in identifying areas where noise levels exceed safe limits and implementing measures to protect workers' hearing.
- ***Music and Audio Engineering*:** Understanding sound intensity is vital in music and audio engineering to achieve the desired loudness and balance in recordings and live performances. It helps in optimizing sound systems and designing acoustically suitable spaces for concerts and recording studios.
- ***Audiology and Hearing Aids*:** Sound intensity measurements are

employed in audiology to diagnose hearing impairments and fit hearing aids. Audiologists use sound intensity data to determine the appropriate amplification required to address specific hearing loss frequencies (Dillon & Storey, 1998).

- ***Traffic and Transportation Planning*:** Sound intensity measurements aid in traffic and transportation planning, particularly for evaluating road and railway noise levels. This data helps in designing quieter transportation infrastructure and reducing noise pollution in residential areas.
- ***Aviation and Aircraft Noise*:** In aviation, sound intensity measurements are essential for assessing aircraft noise levels during takeoff and landing. This data is used to develop quieter aircraft designs and evaluate the impact of airport noise on nearby communities.
- ***Environmental Impact Assessments*:** Sound intensity measurements play a role in conducting environmental impact assessments for various projects, such as wind farms, construction projects, and industrial installations. This helps in understanding and mitigating potential noise pollution effects on wildlife and communities.
- ***Entertainment Industry*:** In movie theaters and entertainment venues, sound intensity measurements are used to ensure that audio levels are appropriate and comply with safety guidelines to prevent hearing damage to patrons (Nathan et al., 2017).

3.3.4. Effects of Sound Intensity

High sound intensity levels, particularly in industrial settings and at concerts or events, can have several effects on individuals and the environment (Figure 3.4):

- ***Hearing Damage*:** Prolonged exposure to high-intensity sounds can lead to permanent hearing damage or hearing loss. This is a significant concern for workers in noisy environments and individuals attending loud concerts or music events without hearing protection (Oftedal, 1988).

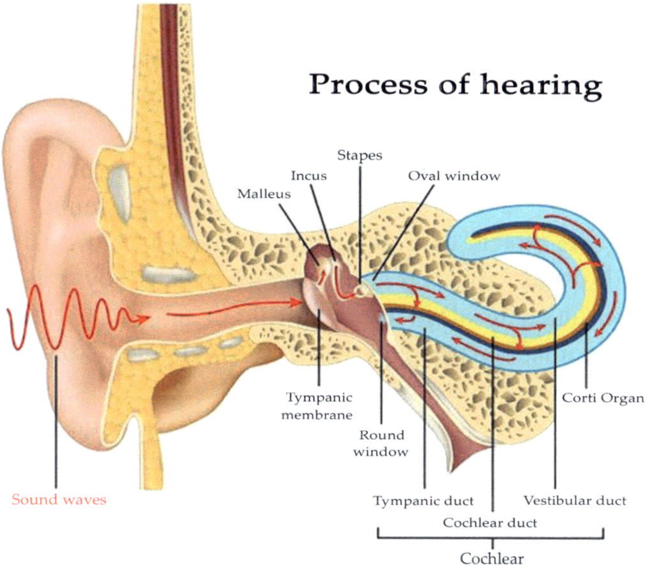

Figure 3.4. Higher sound intensity can damage the hearing of ear (Source: Hitesh Giri, Creative Commons License)

Remember

High sound intensity levels can have adverse effects on human health, such as hearing impairment and other physiological and psychological impacts. It is crucial to limit exposure to excessively loud sounds to protect hearing.

- ***Annoyance and Sleep Disturbance*:** Excessive noise levels can cause annoyance and sleep disturbances in residential areas, leading to reduced quality of life and potential health issues.
- ***Interference with Communication*:** High-intensity sounds can interfere with communication, making it difficult for people to understand each other or hear important warnings and announcements.
- ***Impact on Wildlife*:** Noise pollution can have adverse effects on wildlife, causing behavioral changes, habitat disruption, and interference with mating and hunting patterns.
- ***Structural Damage*:** Extremely high-intensity sounds can cause structural damage to buildings and other structures, particularly when the frequency of the sound matches the natural frequency of the material.

Sound intensity is a fundamental aspect of sound waves that quantifies the amount of sound energy transmitted per unit area. The decibel scale, based on a logarithmic relationship, allows for efficient representation of a wide range of sound intensities. Sound intensity measurements find applications in diverse fields, including environmental monitoring, industrial noise control, music, audiology, and transportation planning. Additionally, understanding the effects of high sound intensity is critical in mitigating potential health and environmental hazards caused by noise pollution. Overall, sound intensity plays a pivotal role in shaping various aspects of our modern society and daily experiences (Asfiati et al., 2020).

3.4. NORMAL MODES OF A STANDING SOUND WAVE

Standing waves are fascinating phenomena that occur when two waves of the same frequency and amplitude traveling in opposite directions interfere with each other. In this section, we will explore the concept of standing waves, the formation of nodes and antinodes, the explanation of normal modes and harmonics, and practical examples and observations of standing sound waves. A standing wave, also known as a stationary wave, is a wave pattern that remains in a fixed position rather than propagating through space. It is the result of the superposition of two waves traveling in opposite directions with the same frequency, wavelength, and amplitude. When these waves interfere constructively and destructively, they create regions of maximum and minimum displacement, leading to the formation of nodes and antinodes (Figure 3.5).

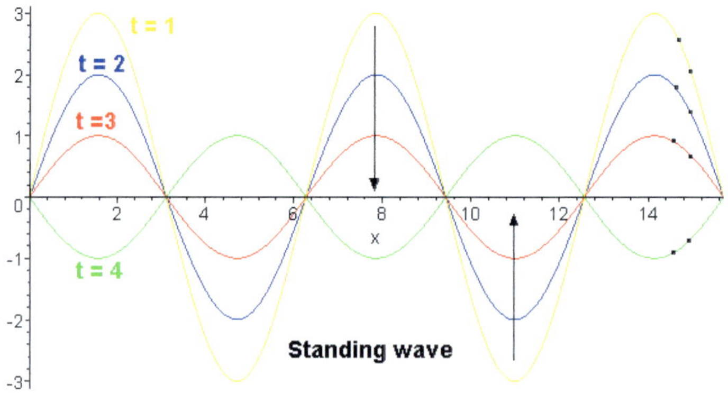

Figure 3.5. Representation of the standing waves (Source: Ualberta, Creative Commons License)

The defining characteristic of standing waves is that certain points along the wave pattern appear to be stationary, while other points exhibit regular oscillations. This results in distinct patterns of vibration that remain fixed in space (Russell et al., 2009).

3.4.1. Formation and Concept of Nodes and Antinodes

Nodes and antinodes are essential components of standing waves and play a crucial role in shaping their unique characteristics:

- ***Nodes*:** Nodes are points along a standing wave where the amplitude of the wave is always zero. At these points, the interfering waves are precisely out of phase, leading to destructive interference. Consequently, the particles at the nodes do not experience any displacement, remaining stationary throughout the wave's motion.
- ***Antinodes*:** Antinodes, on the other hand, are points along the standing wave where the amplitude is at its maximum. At these points, the interfering waves are in phase, leading to constructive interference. As a result, the particles at the antinodes experience the maximum displacement during the wave's motion.

The pattern of nodes and antinodes in a standing wave is constant and does not propagate through space (Figure 3.6). The number of nodes and antinodes determines the mode of the standing wave and is related to its frequency and wavelength (Bulavin et al., 1995).

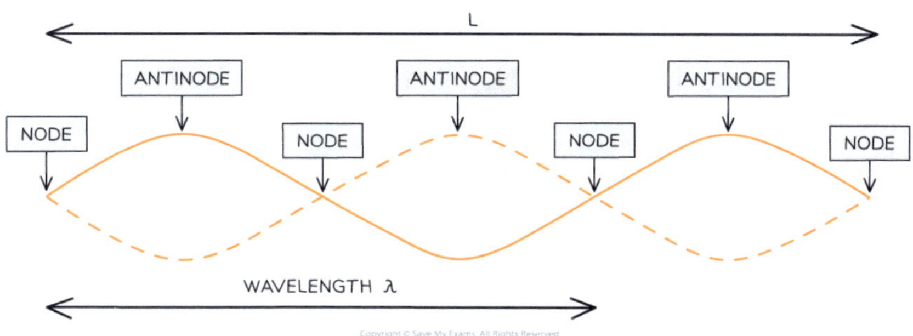

Figure 3.6. Representation of nodes and antinodes (Source: Ashika, Creative Commons License).

3.4.2. Explanation of Normal Modes and Harmonics

3.4.2.1. Normal Modes

Normal modes refer to the different patterns of vibration that can occur in a standing wave. Each normal mode corresponds to a specific frequency of vibration, and the number of nodes and antinodes in the wave determines the mode.

The simplest and fundamental mode of vibration in a standing wave is called the first harmonic or the fundamental mode. It has one antinode at its midpoint and two nodes at its ends. As the frequency of vibration increases, additional nodes and antinodes are formed, giving rise to higher-order harmonics or overtones. These higher harmonics have frequencies that are integer multiples of the fundamental frequency. For example, the second harmonic has two antinodes and three nodes, the third harmonic has three antinodes and four nodes, and so on. Each higher harmonic has a more complex pattern of nodes and antinodes and corresponds to a higher frequency (Pierce, 1965).

3.4.2.2. Harmonics

Harmonics are the individual frequencies present in a complex waveform, and they are integer multiples of the fundamental frequency. In the context of standing waves, harmonics refer to the higher modes of vibration beyond the fundamental mode. The fundamental frequency corresponds to the first harmonic, and subsequent harmonics represent the higher-order modes with increasing frequencies. The fundamental frequency and its harmonics collectively form a harmonic series. Each harmonic in the series has a specific amplitude, and together they contribute to the overall shape of the standing wave (Jaafar et al., 2017).

3.4.3. Practical Examples and Observations

3.4.3.1. Musical Instruments

Standing waves and their harmonics play a fundamental role in producing the distinct sounds of musical instruments (Figure 3.7). For example:

- In a string instrument like a guitar or violin, the length and tension of the string determine the fundamental frequency, while the higher harmonics are responsible for the timbre and tone of the sound produced.

- In wind instruments like flutes and clarinets, the standing waves in the air column inside the instrument create different harmonics, which are controlled by the player's fingerings or the opening of the instrument's holes.
- In percussion instruments like drums and xylophones, the standing waves on the surface of the vibrating material produce different frequencies and pitches (Mustofa et al., 2021).

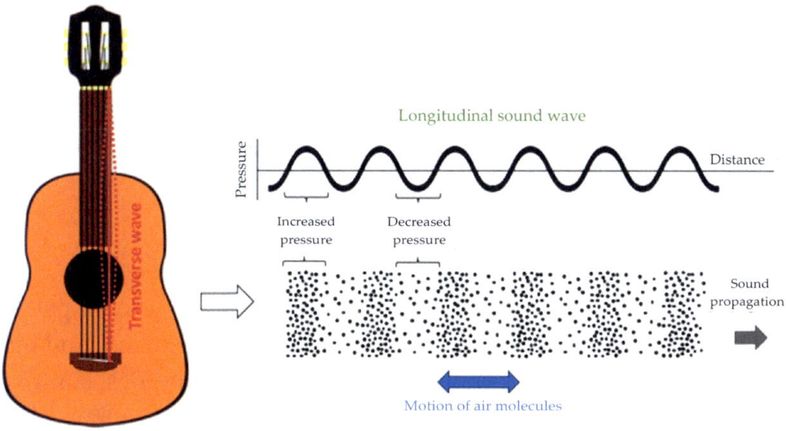

Figure 3.7. The different parts of a musical instrument that work together to amplify sound (Source: Katie Gutierrez, Creative Commons License)

3.4.3.2. Organ Pipes

Organ pipes operate based on the principle of standing sound waves. When air is blown across the top of an open pipe or through a narrow slit in a closed pipe, standing waves are established inside the pipe. The fundamental frequency and harmonics of the standing waves determine the pitch of the note produced by the organ pipe.

3.4.3.3. Room Acoustics

In rooms and concert halls, standing waves can influence the acoustics of the space. When sound waves reflect off the walls, floor, and ceiling, they can interfere constructively or destructively, creating areas of reinforced sound (higher amplitude) or reduced sound (lower amplitude) known as acoustic modes. This can result in uneven sound distribution within the room.

Sound engineers and architects take these standing wave patterns into account when designing acoustically optimized spaces to ensure even sound distribution and minimize undesirable acoustic effects, such as echo and resonance (Lokki et al., 2013).

3.4.3.4. Resonance

Standing waves are closely associated with the concept of resonance, where a vibrating system responds strongly to external forces that match its natural frequency. Resonance can lead to amplification of sound and is exploited in various musical instruments and sound systems.

For example, when a vocalist sings a note that matches the natural frequency of a glass, the glass can resonate, and the sound wave will create a standing wave pattern within the glass, causing it to vibrate and produce a distinct pitch.

3.4.3.5. Vibrating Strings and Chords

Musical instruments like pianos and guitars utilize standing wave patterns in vibrating strings to produce different notes. The length, tension, and mass of the strings determine the fundamental frequency and harmonics, contributing to the variety of sounds and pitches produced by these instruments (Rogers, 2004).

3.4.3.6. Seismic Waves

Standing waves can also occur in the Earth's crust during seismic events. When seismic waves traveling through the Earth encounter areas with similar frequencies, standing waves can form, leading to localized areas of increased ground motion.

In conclusion, normal modes of standing sound waves are fascinating phenomena that result from the interference of waves traveling in opposite directions. Nodes and antinodes characterize standing waves, and the number of nodes and antinodes determines the normal mode or harmonic of the standing wave. Standing waves are found in musical instruments, organ pipes, room acoustics, and even seismic events, contributing to the production of distinct sounds and various acoustic effects. Understanding standing waves and their harmonics is crucial in fields like music, acoustics, and seismology, enriching our knowledge of wave behavior and its applications in practical settings (Kuznetsov, 2011).

3.5. SOURCES OF MUSICAL SOUND

Musical sound is a delightful form of auditory expression that has been an integral part of human culture for thousands of years. It is created by vibrating objects that produce sound waves, which are then perceived by our ears as music. Understanding the sources of musical sound and how pitch is related to these sources is essential in comprehending the mechanics of music production. In this section, we will explore the introduction to musical sound and pitch and how these elements form the foundation of the world of music.

Musical sound is characterized by two primary elements: pitch and timbre. Pitch refers to the highness or lowness of a sound and is directly related to the frequency of the sound waves produced by a vibrating object. Higher-frequency waves are perceived as higher-pitched sounds, while lower-frequency waves are perceived as lower-pitched sounds. In musical terms, pitch is used to define the specific notes that make up melodies and harmonies.

In the context of music, the pitch of a sound is typically represented using the Western musical notation system, where notes are named using the letters A to G, followed by a series of sharps (#) or flats (b) to indicate alterations. The pitch of a note is further divided into octaves, where each higher octave represents a doubling of the frequency of the fundamental note (Di Stefano, 2022).

3.5.1. Vibrating Systems

At the core of musical sound production lies the concept of vibrating systems. A vibrating system is any object or structure that can oscillate back and forth around an equilibrium position. When a vibrating system is set into motion, it generates sound waves that propagate through the surrounding medium, such as air or water.

The primary vibrating systems used in music are strings, air columns, and membranes.

- ***Strings*:** String instruments, such as guitars, violins, and pianos, produce sound by plucking, bowing, or striking strings. The strings vibrate and create sound waves in the air, which then reach our ears and are perceived as musical notes. The pitch of the note produced by a vibrating string is determined by factors such as the length, tension, and mass of the string.

- ***Air Columns*:** Wind instruments, like flutes, clarinets, and trumpets, use vibrating air columns to produce sound. When a player blows air into the instrument, it sets the air column inside the instrument into vibration. By changing the length of the air column (by covering or uncovering holes or using valves), different pitches are produced.
- ***Membranes*:** Membrane instruments, such as drums and timpani, produce sound by striking a stretched membrane. The membrane vibrates, creating sound waves in the surrounding air, and the pitch is determined by factors like the tension and size of the membrane (Desvillettes & Lorenzani, 2012).

3.5.2. Classification of Sound Production

Musical instruments are categorized based on the mechanism they use to produce sound. The main categories of musical instruments are:

1. ***String Instruments*:** As mentioned earlier, string instruments produce sound by vibrating strings. They can be further classified into two subcategories:
    - *Plucked String Instruments*: Instruments like the guitar and harp, where the strings are plucked to produce sound.
    - *Bowed String Instruments*: Instruments like the violin and cello, where the strings are bowed to create sound (Demoucron & Causse, 2007).
2. ***Wind Instruments*:** Wind instruments produce sound by vibrating air columns. They are further divided into two categories:
    - *Woodwind Instruments*: Instruments like the flute, clarinet, and saxophone, where the player blows air across a sharp edge or through a reed to create vibrations.
    - *Brass Instruments*: Instruments like the trumpet, trombone, and tuba, where the player's lips vibrate against a mouthpiece to produce sound.
3. ***Percussion Instruments*:** Percussion instruments produce sound by striking or shaking them. They can be further classified into:
    - *Membrane Percussion Instruments*: Instruments like the drum and timpani, where a stretched membrane is struck to produce sound.

- *Non-Membrane Percussion Instruments*: Instruments like cymbals and tambourines, where the sound is produced by striking or shaking solid materials.

4. ***Keyboard Instruments*:** Keyboard instruments produce sound by striking strings or air columns using keys. Examples include the piano and the pipe organ (Godøy et al., 2005).

3.5.3. Role of Resonance

Resonance is a critical concept in music, particularly in musical instruments. Resonance occurs when a vibrating system is exposed to an external force at its natural frequency, causing it to vibrate with maximum amplitude. This phenomenon amplifies the sound produced by the vibrating system (Figure 3.8).

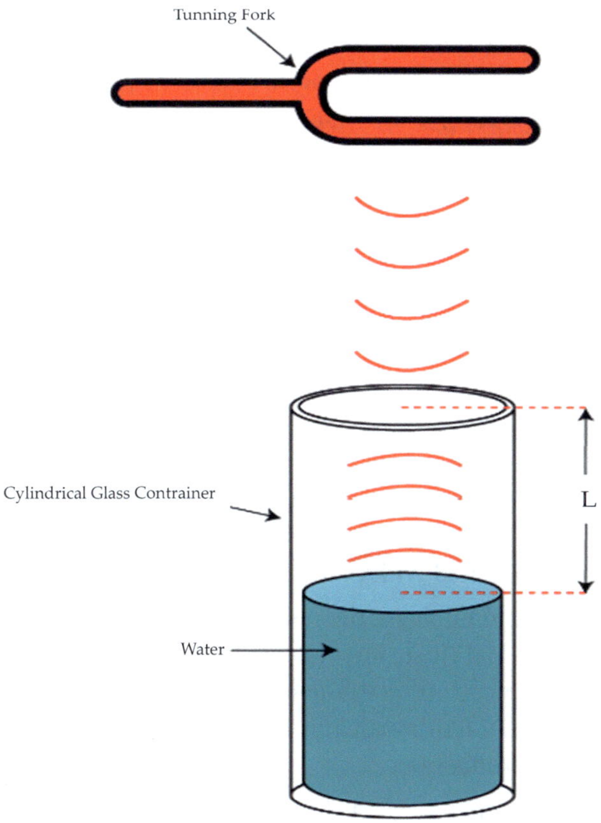

Figure 3.8. Resonance with sound waves (Sound: Education, Creative Commons License).

In musical instruments, resonance plays a vital role in enhancing the sound produced:

- ***Resonance in String Instruments*:** When a string is plucked or bowed, it vibrates at its natural frequency, producing the fundamental pitch. However, the string also produces harmonics (overtones) at multiples of the fundamental frequency. The body of the instrument, such as the hollow chamber of a guitar or violin, acts as a resonator, amplifying these harmonics and enriching the timbre of the sound (Lewandowski & Foltynowicz, 2022).
  - ***Resonance in Wind Instruments*:** In wind instruments, resonance occurs in the air column or the enclosed space inside the instrument. By changing the length of the air column or using valves, the player can create resonance at different frequencies, producing different pitches. The shape and size of the instrument's body also contribute to the resonant frequencies and timbre of the sound produced.
  - ***Resonance in Percussion Instruments*:** In percussion instruments, resonance is achieved through the material and shape of the instrument. For example, the body of a drum or the material of a cymbal is designed to vibrate sympathetically with the striking motion, amplifying the sound and enhancing its tonal qualities.
  - ***Resonance in Keyboard Instruments*:** In keyboard instruments like the piano, resonance occurs due to the interaction of the vibrating strings with the soundboard and the wooden frame. When a key is pressed, a hammer strikes the string, causing it to vibrate. The soundboard amplifies the vibrations and produces the rich and resonant sound associated with pianos.

Musical sound and pitch are the result of vibrating systems that produce sound waves. Musical instruments, such as string instruments, wind instruments, and percussion instruments, utilize vibrating systems to create sound. The classification of musical instruments is based on the mechanism they use for sound production.

Additionally, resonance is a critical aspect of musical instruments, amplifying the sound and contributing to the timbre and richness of the music produced. Understanding the sources of musical sound and the role of resonance allows musicians and composers to create a wide array of sounds and textures, enriching the world of music with its diversity and beauty (Wohltjen & Dessy, 1979).

3.6. BEATS

In the realm of sound, beats are a fascinating phenomenon that arises when two sound waves of slightly different frequencies interfere with each other. Beats are characterized by periodic variations in amplitude, resulting in a pulsating sound. This intriguing acoustic effect has significant implications in music, engineering, and various other fields. In this section, we will explore the definition and concept of beats, the formation and characteristics of beats, the mathematical expression of beat frequency, and the applications of beats in music and engineering.

3.6.1. Formation and Characteristics of Beats

The formation of beats is a consequence of the superposition of two sound waves with slightly different frequencies. When two sound waves of slightly different frequencies travel through the same medium and overlap, they interfere constructively and destructively. Constructive interference occurs when the peaks of the two waves align, reinforcing each other and creating regions of increased amplitude. Destructive interference occurs when the peaks of one wave align with the troughs of the other, resulting in regions of decreased amplitude (Figure 3.9).

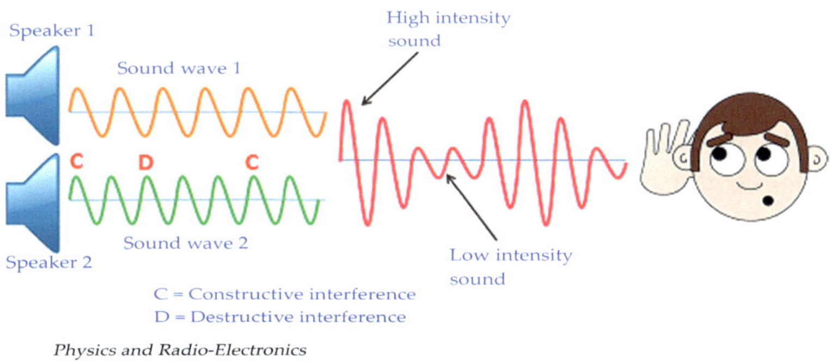

Figure 3.9. Schematic of Beats (Source: Physics and Radio-Electronics, Creative Commons License).

The result of this interference is a periodic variation in the amplitude of the combined wave. These periodic variations produce the sensation of pulsations or beats in the perceived sound. The number of beats per second, known as the beat frequency, is equal to the difference in frequency between the two sound waves.

For example, if two sound waves with frequencies of f1 and f2 (where f1 > f2) interfere, the beat frequency (f_beat) is given by:

f_beat = |f1 – f2|

The beat frequency is the rate at which the amplitude of the combined wave oscillates between maximum and minimum values (Zhang et al., 2015).

3.6.2. Mathematical Expression of Beat Frequency

The mathematical expression of beat frequency can be derived using trigonometric identities. When two waves of different frequencies (A_1 * sin(2πf_1t) and A_2 * sin(2πf_2t)) are added together, the resulting wave is:

A_total * sin(2πf_1t) + A_total * sin(2πf_2t)

Where A_total represents the combined amplitude of the two waves. Using the trigonometric identity for the sum of two sine waves, we get:

A_total * sin(2πf_1t) + A_total * sin(2πf_2t) = 2 * A_total * sin(π(f_1 + f_2) t) * cos(π(f_1 – f_2)t)

The amplitude modulation caused by the cosine term gives rise to the beats. The beat frequency is given by the difference in frequencies:

f_beat = |f_1 – f_2|

3.6.3. Applications of Beats

Beats find various applications in music, engineering, and other fields. Some notable applications include:

- ***Music Tuning and Intonation*:** Beats play a crucial role in tuning musical instruments. When two notes are played together, the presence of beats indicates that they are not perfectly in tune. Musicians can use this phenomenon to adjust the pitch of one of the notes slightly to eliminate the beats and achieve accurate intonation.
- ***Beat Frequencies in Music*:** In music composition, composers can intentionally use beat frequencies to create interesting effects. By combining musical tones with slightly different frequencies, they can produce beats with specific rhythmic patterns, enhancing

the overall musical experience (Cheng et al., 2022).

- ***Psychoacoustics and Hearing Tests*:** Beats are also employed in psychoacoustics research and hearing tests. They help in studying the human auditory system's perception and sensitivity to differences in sound frequencies.
- ***Engineering Applications*:** In engineering, beats are utilized in various applications:
  - *Mechanical Engineering:* In mechanical systems with rotating components, beat frequencies can arise due to small variations in rotational speeds. Detecting and analyzing these beat frequencies can provide valuable information about the condition and health of the machinery.
  - *Electrical Engineering:* In electrical engineering, beat frequencies can occur in electronic circuits with multiple signals or oscillators. Understanding these frequencies is essential for designing and troubleshooting circuits.
  - *Frequency Measurement:* Beat frequencies are used to measure the frequency of a signal accurately. By adjusting the frequency of a known source until beats are eliminated with an unknown frequency source, the unknown frequency can be determined.
- ***Radio Communication:*** In radio communication, beat frequencies are utilized in heterodyning, a process that involves mixing two radio-frequency signals to produce a lower frequency, which is easier to process and detect. This technique is used in superheterodyne receivers, which are common in radio and television receivers.
- ***Astronomical Applications:*** In astronomy, the phenomenon of beats has been used to study binary star systems. When two stars orbit around their common center of mass, the Doppler effect causes their spectral lines to shift periodically, leading to beat frequencies in their observed light spectra.

Beats are a fascinating acoustic phenomenon that occurs when two sound waves of slightly different frequencies interfere with each other, resulting in periodic variations in amplitude. Beat frequencies find applications in various fields, including music tuning, composition, psychoacoustics, engineering, and radio communication. Understanding the concept of beats and their mathematical expression allows us to utilize this intriguing phenomenon in practical settings, enriching music, and advancing engineering applications (Savini et al., 2016).

Did You Know?

Beats have practical applications in music, where musicians use them to tune their instruments. When two slightly different frequencies are played simultaneously, beats are produced, indicating that the instruments are not in tune. Musicians can use this phenomenon to adjust their instruments to achieve harmony.

3.7. THE DOPPLER EFFECT

The Doppler Effect is a fascinating phenomenon observed when there is relative motion between a source of waves and an observer. It results in a change in the frequency or wavelength of the waves as perceived by the observer. The Doppler Effect is prevalent in various aspects of physics, particularly in acoustics and optics, and it has significant real-life applications. In this section, we will delve into the understanding of the Doppler Effect, its application to sound waves, the real-life applications of the Doppler Effect, and the phenomena of redshift and blueshift (Figure 3.10).

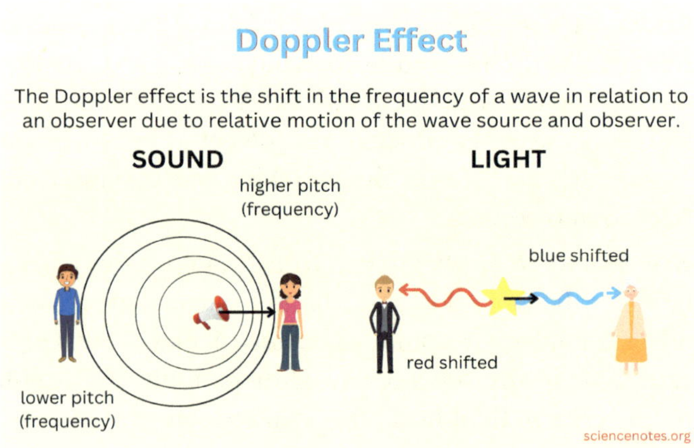

Figure 3.10. Schematic of the Doppler Effect (Source: Science Notes, Creative Commons License).

The Doppler Effect was first described by Christian Doppler, an Austrian mathematician and physicist, in 1842. He postulated that the frequency of waves, such as sound or light waves, would appear different to an observer depending on whether the source and the observer were approaching each other or moving away from each other.

The Doppler Effect is based on the principle that when a source of waves is moving towards an observer, the observer encounters more wave crests per unit time, leading to a higher perceived frequency. Conversely, when the source is moving away from the observer, the observer encounters fewer wave crests per unit time, resulting in a lower perceived frequency (Kündig, 1963).

3.7.1. Doppler Shift for Sound Waves

In the context of sound waves, the Doppler Effect is most commonly experienced with moving sound sources, such as vehicles with sirens or horns, and moving observers, like a person standing on the sidewalk. The key factors that determine the magnitude of the Doppler shift in sound waves are the relative velocity between the source and the observer and the speed of sound in the medium through which the sound waves propagate (usually air).

1. **Moving Sound Source Towards a Stationary Observer:**

When a sound source approaches a stationary observer, the sound waves emitted by the source are compressed, leading to a higher perceived frequency. As the source gets closer, the observer hears a higher-pitched sound. This is commonly experienced when an emergency vehicle with a siren approaches a person standing on the roadside (Rosenblum et al., 1987).

2. **Moving Sound Source Away from a Stationary Observer:**

When a sound source moves away from a stationary observer, the sound waves are stretched, resulting in a lower perceived frequency. As the source moves farther, the observer hears a lower-pitched sound. An example of this is the sound of a train horn fading away as the train moves away from a person on the platform.

3. **Stationary Sound Source with a Moving Observer:**

Conversely, if the sound source is stationary, and the observer is moving towards or away from the source, the Doppler Effect still applies. The relative motion between the observer and the source leads to changes in the perceived frequency.

4. **Formula for the Doppler Shift in Sound Waves:**

The formula to calculate the Doppler shift in sound waves is as follows:

f_observed = f_source * (v_sound + v_observer) / (v_sound + v_source)

Where:

- f_observed is the observed frequency of the sound by the observer,
- f_source is the frequency of the sound emitted by the source,
- v_sound is the speed of sound in the medium (approximately 343 meters per second in air at room temperature),
- v_observer is the velocity of the observer relative to the medium (positive if the observer is moving towards the source, negative if moving away),
- v_source is the velocity of the source relative to the medium (positive if the source is moving towards the observer, negative if moving away) (Mahajan & Gaskill, 1974).

3.7.2. Applications of Doppler Effect in Real Life

The Doppler Effect has numerous practical applications in various fields. Some of the significant real-life applications include:

- ***Doppler Radar*:** Doppler radar is widely used in weather monitoring and forecasting. It measures the Doppler shift of radio waves reflected off precipitation particles, such as raindrops or snowflakes, to detect their motion and estimate wind speed and direction in severe weather conditions.
- ***Doppler Ultrasound*:** In medical imaging, Doppler ultrasound is used to measure blood flow velocity and detect abnormalities in blood vessels. By analyzing the Doppler shift in ultrasound waves reflected from moving red blood cells, doctors can assess blood flow and diagnose vascular conditions (Meimoun & Tribouilloy, 2008).
- ***Police Radar Guns*:** Police officers use Doppler radar guns to measure the speed of moving vehicles accurately. The radar gun emits microwaves that bounce off the moving vehicle and return to the gun. The frequency shift of the returning waves allows officers to calculate the vehicle's speed.
- ***Astronomy*:** In astronomy, the Doppler Effect is crucial for studying the motion of celestial objects. The redshift and blueshift

phenomena help astronomers determine the velocity and distance of stars, galaxies, and other cosmic objects.

- ***Sonar Systems***: Sonar systems on ships and submarines use the Doppler Effect to measure the velocity of underwater objects, such as other vessels or marine life. By analyzing the frequency shift of sound waves reflected from these objects, the system can estimate their motion.
- ***Traffic Flow Analysis***: The Doppler Effect can be used to analyze traffic flow on highways and roadways.

By analyzing the frequency shift of radio waves reflected off moving vehicles, traffic engineers can estimate traffic density and speed, helping to manage traffic flow efficiently (Kashinath et al., 2021).

3.7.3. Redshift and Blueshift Phenomena

In astrophysics, redshift and blueshift are phenomena resulting from the Doppler Effect when observing light from distant celestial objects.

- ***Redshift***: Redshift occurs when a light source is moving away from an observer. As the object moves away, the wavelengths of the light it emits are stretched, resulting in a shift towards the red end of the electromagnetic spectrum. Redshift is a fundamental concept in cosmology and is used to estimate the expansion rate of the universe, as demonstrated by Edwin Hubble's discovery of the redshift of galaxies, providing evidence for the expanding universe.
- ***Blueshift***: Blueshift, on the other hand, occurs when a light source is moving towards an observer. As the object moves closer, the wavelengths of the light it emits are compressed, leading to a shift towards the blue end of the electromagnetic spectrum. Blueshift is also essential in astronomy and has been used to study objects like stars, planets, and galaxies with significant radial velocities.

The Doppler Effect is a fascinating phenomenon that describes the change in frequency or wavelength of waves due to relative motion between the source and the observer. The effect is prominent in sound waves and has numerous real-life applications, such as Doppler radar, Doppler ultrasound, police radar guns, and traffic flow analysis. In astronomy, the Doppler Effect leads to the phenomena of redshift and blueshift, providing valuable information about the motion and distance of celestial objects. Overall, the Doppler Effect is a fundamental concept with broad implications in various fields, enriching our understanding of wave behavior and facilitating practical applications in diverse areas of science and technology (Phuong et al., 2022).

3.8. SHOCK WAVES

Shock waves are fascinating and powerful phenomena that occur when an object moves through a medium faster than the speed of sound in that medium. They are characterized by a rapid and intense increase in pressure and temperature, resulting in a distinctive front called the shock front. Shock waves have significant implications in various fields, including aerospace engineering, meteorology, and medicine. In this section, we will introduce the concept of shock waves, explore their formation and properties, compare supersonic and subsonic speeds, and discuss practical applications and hazards associated with shock waves (Figure 3.11).

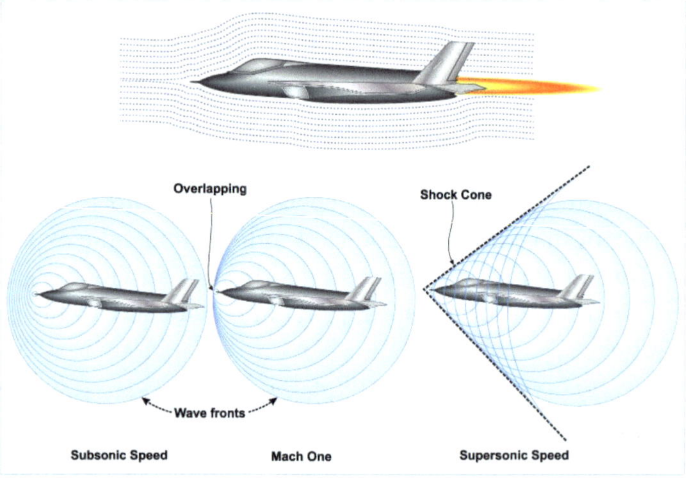

Figure 3.11. How a supersonic aircraft creates shock waves (Source: Fouad A. Saad, Creative Commons License).

When an object travels through a medium, such as air or water, it creates pressure waves that propagate outward from the object at the speed of sound in that medium. In subsonic speeds (speeds slower than the speed of sound), the pressure waves disperse smoothly without any abrupt changes. However, when an object moves at or above the speed of sound, the pressure waves start to pile up and merge, forming a shock front—a narrow region of very high pressure and temperature gradients (Hartmann et al., 2013).

3.8.1. Formation and Properties of Shock Waves

The formation of shock waves is a consequence of the compressibility of the medium and the supersonic motion of the object. When an object moves at or above the speed of sound, it generates a series of pressure waves in front of it. These waves combine and amplify, leading to the creation of a shock front.

Shock waves can be categorized into two main types: normal shock waves and oblique shock waves.

- ***Normal Shock Waves*:** Normal shock waves occur when an object moves directly perpendicular to the direction of the pressure waves it generates. The shock front is perpendicular to the flow direction. Normal shocks are common in various engineering applications, such as in supersonic aircraft and high-speed projectiles (Culick & Rogers, 1983).
- ***Oblique Shock Waves*:** Oblique shock waves occur when an object moves at an angle to the direction of the pressure waves it generates. The shock front is tilted at an angle to the flow direction. Oblique shocks are commonly observed in situations where a flow encounters a solid surface or a change in flow direction.

Shock waves exhibit several unique properties:

- ***Compression*:** Shock waves cause a rapid increase in the pressure of the medium they pass through. The pressure behind the shock front is significantly higher than the pressure ahead of it.
- ***Temperature Increase*:** The compression of the medium by the shock wave leads to a substantial increase in temperature. This rise in temperature is often associated with heating and can have various practical implications.
- ***Energy Dissipation*:** Shock waves carry a considerable amount of kinetic energy, which is rapidly dissipated as they propagate through the medium. This dissipation can be harnessed for certain applications.
- ***Shock Front Propagation*:** Shock waves travel as distinct fronts through the medium, with abrupt changes in pressure and density across the front (Ellzey et al., 1995).

3.8.2. Supersonic vs. Subsonic Speeds

The terms "supersonic" and "subsonic" refer to the speed of an object relative to the speed of sound in the surrounding medium.

- ***Supersonic Speeds*****:** When an object travels at a speed higher than the speed of sound in the medium, it is said to be moving at supersonic speeds. In this regime, shock waves are generated, as described earlier. Supersonic flight has been achieved in various aircraft, including fighter jets and supersonic passenger planes like the Concorde.
- ***Subsonic Speeds*****:** Subsonic speeds refer to object velocities lower than the speed of sound in the medium. In this regime, pressure waves propagate smoothly without forming shock fronts.

The transition from subsonic to supersonic speeds is a critical aerodynamic consideration. At the critical speed called "Mach 1," the object reaches the speed of sound, and shock waves start to form (Nagata et al., 2016).

3.8.3. Practical Applications and Hazards of Shock Waves

Shock waves have numerous practical applications in diverse fields, and their effects can also present hazards in certain situations:

- ***Aerospace Engineering*****:** In the field of aerospace engineering, the study of shock waves is essential for designing supersonic and hypersonic aircraft. Understanding the formation and properties of shock waves helps in optimizing aerodynamic designs and minimizing drag and other undesirable effects associated with supersonic flight (Kozelkov et al., 2022).
- ***Supersonic Propulsion*****:** Supersonic and hypersonic engines rely on the controlled formation and expansion of shock waves to increase the velocity of the exhaust gases and achieve higher thrust.
- ***Shock Wave Lithotripsy*****:** In medicine, shock waves are used in a technique called shock wave lithotripsy to break kidney stones and gallstones into smaller fragments, facilitating their natural expulsion from the body (Streem, 1997).
- ***Sonic Boom*****:** When an aircraft moves at supersonic speeds, it generates a shock wave that propagates outward and is heard on the ground as a loud and sudden noise called a sonic boom.

Sonic booms can be a source of annoyance and are subject to regulations in populated areas.

- ***Explosive Shock Waves*:** In explosives engineering, shock waves are used to initiate controlled explosions for mining, construction, and military purposes.
- ***Meteorology*:** In meteorology, shock waves associated with thunderstorms, known as "shock diamonds" or "Mach diamonds," can be observed in the aftermath of lightning.
- ***Astronomy*:** Shock waves play a role in astrophysical phenomena such as supernova explosions, where they accelerate and heat up surrounding materials (Zirakashvili & Aharonian, 2007).

Hazards of Shock Waves:

- *Sonic Boom Hazards:* Sonic booms produced by supersonic aircraft can lead to structural damage to buildings, disturb wildlife, and cause annoyance and discomfort to people on the ground (Chernyshev et al., 2022).
- *Explosion Hazards:* Uncontrolled shock waves from explosions can cause severe damage to structures and pose risks to human safety (Medici et al., 2014).
- *Meteorological Hazards:* Severe thunderstorms with strong shock waves can cause damage to property, vehicles, and infrastructure.

Shock waves are powerful phenomena that occur when an object moves at or above the speed of sound in a medium. They are characterized by a rapid increase in pressure and temperature and are found in various applications, including aerospace engineering, medical treatments, and explosive engineering. Understanding the formation and properties of shock waves is crucial for optimizing aircraft design, propulsion systems, and medical procedures. However, shock waves can also present hazards, such as sonic booms and explosion risks. As we continue to explore and harness the power of shock waves, it is essential to balance their benefits with the potential risks and take appropriate measures to mitigate any adverse effects (Ren et al., 2023).

Activity 3.1

Imagine you are designing a concert hall for a world-famous orchestra. How would you utilize the concept of sound intensity to ensure optimal acoustics within the hall? Additionally, how would you address the issue of potential standing waves to create a pleasant auditory experience for the audience and musicians?

SUMMARY

This chapter delves into the fascinating world of sound waves, covering essential topics such as the speed of sound, sound intensity, normal modes of standing sound waves, sources of musical sound, beats, the Doppler Effect, and shock waves. It explores the characteristics and behavior of sound waves as they propagate through various media, including air and water. The speed of sound is analyzed in relation to factors affecting its velocity, such as temperature and material properties. Sound intensity is discussed, along with its measurement in decibels and its real-world effects. The chapter also explores the phenomenon of normal modes and harmonics in standing sound waves, which give rise to musical tones in various instruments. It further examines how beats occur when two sound waves interfere and create periodic variations in amplitude. The Doppler Effect is introduced, highlighting how the motion of a sound source relative to an observer can alter the perceived frequency of sound waves. Lastly, the chapter discusses shock waves that form when an object travels faster than the speed of sound, examining their properties, applications in engineering and medicine, and potential hazards like sonic booms. Overall, this chapter provides a comprehensive and fascinating exploration of sound-related phenomena and their practical significance in various fields.

REVIEW QUESTIONS

1. How is the speed of sound determined, and what are the factors that can affect its speed?
2. Explain the concept of sound intensity and how it is measured. Discuss the applications and effects of sound intensity in real-world scenarios.
3. What are the normal modes of a standing sound wave, and what is the significance of nodes and antinodes in these modes?

4. Describe the sources of musical sound, focusing on vibrating systems and their classification in sound production.
5. Define beats and discuss their formation and characteristics. How can the beat frequency be mathematically expressed?
6. What is the Doppler Effect, and how does it apply to sound waves? Provide examples of the Doppler Effect in real-life situations.
7. Differentiate between redshift and blueshift phenomena in the context of the Doppler Effect.
8. Discuss shock waves, their formation, and their properties. What are the differences between supersonic and subsonic speeds?
9. Explore the practical applications and potential hazards associated with shock waves.
10. Summarize the key concepts covered in the chapter on sound waves.

REFERENCES

1. Akins, J. (2006). Facilities used for introductory electronic music: A survey of music production and technology programs. *MEIEA Journal*, *6*(1), 107–127.
2. Aksenov, S. P., & Kuznetsov, G. N. (2021). Determination of interference invariants in a deep-water waveguide by amplitude and phase methods. *Physics of Wave Phenomena*, *29*(1), 81–87.
3. Andrade, M. A., Marzo, A., & Adamowski, J. C. (2020). Acoustic levitation in mid-air: Recent advances, challenges, and future perspectives. *Applied Physics Letters*, *116*(25), 247–250.
4. Asfiati, S., Riky, M. N., & Rajagukguk, J. (2020). Measurement and evaluation of sound intensity at the Medan Railway Station using a sound level meter. In *Journal of Physics: Conference Series*. IOP Publishing. (Vol. 1428, pp. 12063–12064).
5. Awad, N., & Barak, M. (2018). Pre-service science teachers learn a Science, Technology, Engineering and Mathematics (STEM)-oriented program: The case of sound, waves and communication systems. *Eurasia Journal of Mathematics, Science and Technology Education*, *14*(4), 1431–1451.
6. Blake, E. W. (1848). ART. XLII.--A Determination of the General Law according to which Pulses differing in intensity are propagated in Elastic Media, with remarks on the received Theory of the Velocity of Sound. *American Journal of Science and Arts (1820–1879)*, *5*(15), 372.
7. Bulavin, V. V., Gutsev, D. F., & Pavelko, V. I. (1995). The experimental definition of the acoustic standing waves series shapes, formed in the coolant of the primary circuit of VVER-440 type reactor. *Progress in Nuclear Energy*, *29*(3), 153–170.
8. Cheng, C. M., Lin, T. F., Chen, Y. J., & Chen, L. B. (2022). Promoting relaxation using monaural beats with ultralow-frequency inaudible sounds: An empirical case study. *IEEE Access*, *10*, 50807–50818.
9. Chernyshev, S. L., Gorbovskoy, V. S., Kazhan, A. V., & Korunov, A. O. (2022). Re-entry vehicle sonic boom issue: modelling and calculation results in windy atmosphere based on the augmented Burgers equation. *Acta Astronautica*, *194*, 450–460.
10. Cressman, G. P. (1965). Numerical Weather Prediction in Daily Use: The tenth anniversary of calculations for daily weather forecasts finds forecasts more accurate and useful. *Science*, *148*(3668), 319–327.

11. Crocker, M. J., & Arenas, J. P. (2003). Fundamentals of the direct measurement of sound intensity and practical applications. *Acoustical Physics*, *49*(2), 163–175.
12. Culick, F. E. C., & Rogers, T. (1983). The response of normal shocks in diffusers. *AIAA Journal*, *21*(10), 1382–1390.
13. Daigle, G. A. (1979). Effects of atmospheric turbulence on the interference of sound waves above a finite impedance boundary. *The Journal of the Acoustical Society of America*, *65*(1), 45–49.
14. Demoucron, M., & Causse, R. (2007, September). Sound synthesis of bowed string instruments using a gesture based control of a physical model. In *International Symposium on Musical Acoustics* (Vol. 2, pp. 1–1).
15. Desvillettes, L., & Lorenzani, S. (2012). Sound wave resonances in micro-electro-mechanical systems devices vibrating at high frequencies according to the kinetic theory of gases. *Physics of Fluids*, *24*(9), 658–660.
16. Di Stefano, N. (2022). The spatiality of sounds. From sound-source localization to musical spaces. *Aisthesis. Pratiche, linguaggi e saperi dell'estetico*, *15*(1), 173–185.
17. Dillon, H., & Storey, L. (1998). The National Acoustic Laboratories' procedure for selecting the saturation sound pressure level of hearing aids: theoretical derivation. *Ear and Hearing*, *19*(4), 255–266.
18. Ellzey, J. L., Henneke, M. R., Picone, J. M., & Oran, E. S. (1995). The interaction of a shock with a vortex: Shock distortion and the production of acoustic waves. *Physics of Fluids*, *7*(1), 172–184.
19. Elson, A. (1921). Architectural acoustics. *The Musical Quarterly*, *7*(4), 469–482.
20. Godøy, R. I., Haga, E., & Jensenius, A. R. (2005, May). Playing "air instruments": mimicry of sound-producing gestures by novices and experts. In *International Gesture Workshop*. Berlin, Heidelberg: Springer Berlin Heidelberg. (Vol. 2, pp. 256–267).
21. Greenspan, M., & Tschiegg, C. E. (1957). Speed of sound in water by a direct method. *Journal of Research of the National Bureau of Standards*, *59*(4), 249–254.
22. Hartmann, A., Feldhusen, A., & Schröder, W. (2013). On the interaction of shock waves and sound waves in transonic buffet flow. *Physics of Fluids*, *25*(2), 421–425.

23. Heki, K., Otsuka, Y., Choosakul, N., Hemmakorn, N., Komolmis, T., & Maruyama, T. (2006). Detection of ruptures of Andaman fault segments in the 2004 great Sumatra earthquake with coseismic ionospheric disturbances. *Journal of Geophysical Research: Solid Earth*, *111*(9), 5820–5823.
24. Jaafar, R., Daud, A. N. M., Ali, S., & Ayop, S. K. (2017). Three-in-one resonance tube for harmonic series sound wave experiments. *Physics Education*, *52*(4), 045008.
25. Jacobsen, F., & de Bree, H. E. (2005). A comparison of two different sound intensity measurement principles. *The Journal of the Acoustical Society of America*, *118*(3), 1510–1517.
26. Jolly, M. R., Prabhakar, A., Sturzu, B., Hollstein, K., Singh, R., Thomas, S., ... & Shaw, A. (2015). Review of non-destructive testing (NDT) techniques and their applicability to thick walled composites. *Procedia CIRP*, *38*, 129–136.
27. Kaji, S., & Okazaki, T. (1970). Propagation of sound waves through a blade row: II. Analysis based on the acceleration potential method. *Journal of Sound and Vibration*, *11*(3), 355–360.
28. Kashinath, S. A., Mostafa, S. A., Mustapha, A., Mahdin, H., Lim, D., Mahmoud, M. A., ... & Yang, T. J. (2021). Review of data fusion methods for real-time and multi-sensor traffic flow analysis. *IEEE Access*, *9*, 51258–51276.
29. Kasper, L., Vogt, P., & Strohmeyer, C. (2015). Stationary waves in tubes and the speed of sound. *The Physics Teacher*, *53*(1), 52–53.
30. Katz, J. E., Clavijo, R. I., Rizk, P., & Ramasamy, R. (2020). The basic physics of waves, soundwaves, and shockwaves for erectile dysfunction. *Sexual Medicine Reviews*, *8*(1), 100–105.
31. Kozelkov, A. S., Strelets, D. Y., Sokuler, M. S., & Arifullin, R. H. (2022). Application of mathematical modeling to study near-field pressure pulsations of a near-future prototype supersonic business aircraft. *Journal of Aerospace Engineering*, *35*(1), 04021120.
32. Kündig, W. (1963). Measurement of the transverse doppler effect in an accelerated system. *Physical Review*, *129*(6), 2371.
33. Kuwano, S., Namba, S., & Miura, H. (1989). Advantages and disadvantages of A-weighted sound pressure level in relation to subjective impression of environmental noise. *Noise Control Engineering Journal*, *33*(3), 107–115.

34. Kuznetsov, S. V. (2011). Seismic waves and seismic barriers. *Acoustical Physics*, *57*, 420–426.
35. Lewandowski, M., & Foltynowicz, Z. (2022). Selected aspects of resonance properties of different woods and the construction of string instruments. *Drewno. Prace Naukowe. Doniesienia. Komunikaty*, *65*(210), 36981.
36. Li, G., Chen, Y., Chen, W., Liu, J., & He, H. (2022). Local resonance–Helmholtz lattices with simultaneous solid-borne elastic waves and air-borne sound waves attenuation performance. *Applied Acoustics*, *186*, 108450.
37. Lin, W., Xiao, J., Wen, J., & Wang, S. (2022). Identification approach of acoustic cavitation via frequency spectrum of sound pressure wave signals in numerical simulation. *Ultrasonics Sonochemistry*, *90*, 106182.
38. Linder, C. J. (1993). University physics students' conceptualizations of factors affecting the speed of sound propagation. *International Journal of Science Education*, *15*(6), 655–662.
39. Lokki, T., Southern, A., Siltanen, S., & Savioja, L. (2013). Acoustics of Epidaurus–studies with room acoustics modelling methods. *Acta Acustica united with Acustica*, *99*(1), 40–47.
40. Mahajan, V. N., & Gaskill, J. D. (1974). Doppler interpretation of the frequency shifts of light diffracted by sound waves. *Journal of Applied Physics*, *45*(6), 2799–2800.
41. Maia, T. Q., Alvarenga, A. V., Souza, R. M., & Costa-FÚlix, R. P. (2021). Feasibility of reference material certification for speed of sound and attenuation coefficient based on standard tissue-mimicking material. *Ultrasound in Medicine & Biology*, *47*(7), 1904–1919.
42. Medici, E. F., Allen, J. S., & Waite, G. P. (2014). Modeling shock waves generated by explosive volcanic eruptions. *Geophysical Research Letters*, *41*(2), 414–421.
43. Meimoun, P., & Tribouilloy, C. (2008). Non-invasive assessment of coronary flow and coronary flow reserve by transthoracic Doppler echocardiography: a magic tool for the real world. *European Journal of Echocardiography*, *9*(4), 449–457.
44. Mustofa, H. A., Gunawan, G., & Kosim, K. (2021). Validity of documentary video learning media based on local content traditional musical instruments on sound waves. *Jurnal Pijar Mipa*, *16*(5), 576–582.

45. Nagata, T., Nonomura, T., Takahashi, S., Mizuno, Y., & Fukuda, K. (2016). Investigation on subsonic to supersonic flow around a sphere at low Reynolds number of between 50 and 300 by direct numerical simulation. *Physics of Fluids*, *28*(5), 321–323.
46. Nathan, G., Britten, F., & Burnett, J. (2017). Sound Intensity Levels of Volume Settings on Cardiovascular Entertainment Systems in a University Wellness Center. *Recreational Sports Journal*, *41*(1), 20–26.
47. Njeh, C. F., Richards, A., Boivin, C. M., Hans, D., Fuerst, T., & Genant, H. K. (1999). Factors influencing the speed of sound through the proximal phalanges. *Journal of Clinical Densitometry*, *2*(3), 241–250.
48. Oftedal, G. (1988). Noise-induced hearing damage caused by metabolic exhaustion: A mathematical model. *The Journal of the Acoustical Society of America*, *83*(4), 1499–1507.
49. Phuong, L. T., Hoi, B. D., & Yarmohammadi, M. (2022). Optical redshift and blueshift spectra in monolayer β 12-borophene: Inversion symmetry breaking effects. *Physical Review B*, *106*(19), 195409.
50. Pierce, A. D. (1965). Extension of the method of normal modes to sound propagation in an almost-stratified medium. *The Journal of the Acoustical Society of America*, *37*(1), 19–27.
51. Ren, Z., Higuera, P., & Li-Fan Liu, P. (2023). On Tsunami Waves Induced by Atmospheric Pressure Shock Waves After the 2022 Hunga Tonga-Hunga Ha'apai Volcano Eruption. *Journal of Geophysical Research: Oceans*, *128*(4), 5410–5413.
52. Robin, O., & Plante, C. (2022). An illustrated tutorial for logarithmic scales and decibels in acoustics. *The Journal of the Acoustical Society of America*, *152*(5), 2880–2892.
53. Rogers, G. L. (2004). Interdisciplinary lessons in musical acoustics: The science-math-music connection. *Music Educators Journal*, *91*(1), 25–30.
54. Rosenblum, L. D., Carello, C., & Pastore, R. E. (1987). Relative effectiveness of three stimulus variables for locating a moving sound source. *Perception*, *16*(2), 175–186.
55. Russell, D. A., Parker, D. E., & Hughes, R. S. (2009). Analysis of standing sound waves using holographic interferometry. *American Journal of Physics*, *77*(8), 678–682.
56. Saha, S., Singh, A., Bera, P., Kamal, M. N., Dutta, S., Gorian, U., ...

& Sur, S. (2017, October). GPS based smart spy surveillance robotic system using Raspberry Pi for security application and remote sensing. In *2017 8th IEEE Annual Information Technology, Electronics and Mobile Communication Conference (IEMCON)*. IEEE. (Vol. 2, pp. 705–709).

57. Santos, R. S., & Macedo Filho, E. (2006). Sonar doppler as an instrument of deglutition evaluation. *Int Arch Otorhinolaryngol*, *10*(3), 182–91.
58. Savini, G., Brooks, D., Zhu, L., Ellison, B., Auriacombe, O., Bradshaw, T., ... & Yu, G. (2016, September). Recent progress in the Elegant Breadboard supra-THz activities for LOCUS and a view to an astronomy application. In *2016 IEEE 9th UK-Europe-China Workshop on Millimeter Waves and Terahertz Technologies (UCMMT)*. IEEE. (Vol. 2, pp. 1–5).
59. Sawada, Y., & Ohsumi, A. (2000, June). Active noise control of sound wave in a one-dimensional duct. In *Proceedings of the 2000 American Control Conference. ACC (IEEE Cat. No. 00CH36334)*. IEEE. (Vol. 5, pp. 3013–3017).
60. Sitdikova, L. F., & Gimaltdinov, I. K. (2022). Reflection and Refraction of Sound Waves at the Interface Between a Bubbly Liquid and a Porous Medium Saturated with the Bubbly Liquid. *Journal of Applied Mechanics and Technical Physics*, *63*(5), 851–859.
61. Stähli, P., Frenz, M., & Jaeger, M. (2020). Bayesian approach for a robust speed-of-sound reconstruction using pulse-echo ultrasound. *IEEE Transactions on Medical Imaging*, *40*(2), 457–467.
62. Stojanovic, M., & Preisig, J. (2009). Underwater acoustic communication channels: Propagation models and statistical characterization. *IEEE communications magazine*, *47*(1), 84–89.
63. Streem, S. B. (1997). Contemporary clinical practice of shock wave lithotripsy: a reevaluation of contraindications. *The Journal of urology*, *157*(4), 1197–1203.
64. Thol, M., Dubberke, F. H., Baumhögger, E., Vrabec, J., & Span, R. (2017). Speed of sound measurements and fundamental equations of state for octamethyltrisiloxane and decamethyltetrasiloxane. *Journal of Chemical & Engineering Data*, *62*(9), 2633–2648.
65. Tomonaga, S. I. (1950). Remarks on Bloch's method of sound waves applied to many-fermion problems. *Progress of Theoretical Physics*, *5*(4), 544–569.

66. Wittmann, M., Steinberg, R. N., & Redish, E. F. (2003). Understanding and affecting student reasoning about sound waves. *International Journal of Science Education*, *25*(8), 991–1013.
67. Wohltjen, H., & Dessy, R. (1979). Surface acoustic wave probe for chemical analysis. I. Introduction and instrument description. *Analytical chemistry*, *51*(9), 1458–1464.
68. Zhang, Y., Moore, K. J., McFarland, D. M., & Vakakis, A. F. (2015). Targeted energy transfers and passive acoustic wave redirection in a two-dimensional granular network under periodic excitation. *Journal of Applied Physics*, *118*(23), 987–990.
69. Zirakashvili, V. N., & Aharonian, F. (2007). Analytical solutions for energy spectra of electrons accelerated by nonrelativistic shock-waves in shell type supernova remnants. *Astronomy & Astrophysics*, *465*(3), 695–702.

CHAPTER 4

Light Waves

Contents

4.1. INTRODUCTION

Light waves are a type of electromagnetic wave characterized by the presence of oscillating magnetic and electric fields. The waves propagate through the medium of space at a velocity equivalent to the speed of light, which is approximately 299,792 km/s. Furthermore, these waves interact with matter, enabling human beings to perceive and comprehend their surroundings via the faculty of vision. Light waves possess dual characteristics, displaying properties of both waves and particles, specifically exhibiting traits of waves and photons, which are discrete units of energy. The electromagnetic spectrum encompasses the complete continuum of light waves, spanning from radio waves characterized by longer wavelengths and lower frequencies to microwaves, infrared radiation, visible light (the segment perceptible to the human eye), ultraviolet radiation, X-rays, and gamma rays characterized by shorter wavelengths and higher frequencies (Boudiombo et al., 1997).

The perception of colors is attributed to visible light, which encompasses wavelengths spanning approximately 400 to 700 nanometers. When a beam of white light traverses a prism, it experiences dispersion, resulting in the segregation of its constituent colors and the formation of a spectrum ranging from violet to red. The interaction between light waves and matter leads to a range of optical phenomena, including reflection, refraction, diffraction, and interference. These phenomena serve as the foundation for a wide range of practical applications in daily life, encompassing lenses, mirrors, cameras, and fiber-optic communication. Light waves are of utmost importance in the field of astronomy due to their fundamental role as the conveyors of information originating from celestial objects located at significant distances. Telescopes and other astronomical instruments employ light waves as a means to collect data and investigate the cosmos (Rumsey et al., 1951).

Within the realm of quantum mechanics, it is observed that light waves exhibit a characteristic akin to that of particles, specifically photons. Photons, which are particles devoid of mass, possess discrete units of energy and exhibit dual characteristics as both particles and waves. The phenomenon of wave-particle duality in relation to light serves as a fundamental principle within the framework of quantum theory. In summary, light waves are a type of electromagnetic wave that propagates through the vacuum of space, enabling the human visual system to perceive and comprehend the surrounding environment. Photons possess both wave-like and particle-like characteristics, rendering them indispensable in diverse domains such as optics, astronomy, and quantum mechanics. The ongoing investigation into light waves serves to elucidate the enigmatic aspects of the cosmos and contribute to the development of technological innovations that enhance our everyday existence (Bauke et al., 2014).

4.2. PHOTONICS

Photonics can be described as the comprehensive field encompassing the processes of generating, manipulating, transporting, detecting, and utilizing both information and energy carried by light. It is vital to note that the fundamental unit of light in this context is the photon, which operates on a quantum level. Photonics is a field of study that is rooted in the principles of electronics and optics. The origins of photonics can be traced back to ancient times. Exodus 38:8, believed to have been written around 1200 BCE, mentions the presence of "looking glasses" among women. Photonics is anticipated to play a crucial role in the forthcoming century as an integral component of the information revolution. (Meng et al., 2021).

In order to facilitate comprehension and practical application of photonics, it is imperative to possess a fundamental comprehension of the characteristics and behaviors exhibited by light. This course centers its attention on the fundamental principles of photonics, thereby equipping individuals with the necessary knowledge and skills to pursue a promising career as a photonics technician.

There exist ten overarching domains of occupation that are anticipated to experience a growing demand for photonics technicians (Yablonovitch, 2001):

- Medicine-biomedical
- Environmental
- Energy
- Transportation
- Defense
- Public safety
- Aerospace
- Computers
- Manufacturing with photonics and test and analysis
- Communication and information technology noninvasive diagnostic tools.

The field of medicine has experienced substantial advancements in the utilization of photonics devices. The expansion results in increasingly abundant prospects for technicians specializing in biomedical photonics (Kim et al., 2020).

In the realm of environmental science, photonics devices have the capability to remotely assess the presence of pollutants in both our atmospheric and aquatic ecosystems. Photonics devices have the capability to efficiently capture and utilize renewable energy derived from the sun, thereby enhancing the overall energy supply and contributing to the conservation of our finite natural resources. The field of transportation is expected to experience substantial transformations, exemplified by the implementation of the Intelligent Vehicle Highway System. This advanced system offers various features, including guidance, collision avoidance, and continuous engine optimization in response to driving conditions. Furthermore, the defense industry has identified a number of photonics devices capable of effectively neutralizing hostile weaponry (Baba & Nakamura, 2002).

Photonics will play a crucial role in collecting, manipulating, storing, transmitting, and visualizing information during the Information Age. There have been proposals for the utilization of novel optical computers in certain applications, while charged coupled devices (CCD) cameras have been employed to digitize artwork. Compact discs (CDs), digital video devices (DVDs), and various other media are employed to store and retrieve data through laser technology. Fiber optics is utilized for establishing connections between nodes of the Internet or telephone lines. Data can be outputted through the utilization of laser printers. Photonics is experiencing rapid growth, and the career prospects for technicians in this domain are highly promising.

The implementation of photonics plays a crucial role in the automation of manufacturing processes. Fabrication processes are predominantly carried out utilizing industrial lasers, which are employed for the purposes of cutting, welding, trimming, drilling holes, and heat-treating various products. In order to ensure the quality of the product, a comprehensive inspection process is implemented, which involves the utilization of spectroscopy, interferometry, machine vision, and image processing techniques. The increasing sophistication of photonics in the manufacturing sector is anticipated to result in a substantial surge in the demand for proficient photonics technicians (Soukoulis, 2012).

4.3. PROPERTIES OF LIGHT

Light refers to electromagnetic radiation visible to the human eye, typically characterized by its ability to illuminate objects and enable visual perception. The topic at hand has been the subject of extensive scholarly discourse spanning numerous centuries. The sun emits radiant energy in the form of light, while electric lighting serves to illuminate our surroundings during periods of darkness. Numerous other applications of light significantly influence our daily existence. In brief, light can be characterized as a distinct form of electromagnetic radiation (Hashimoto et al., 2007).

The velocity of light, despite its considerable magnitude, is finite. The velocity of light in a vacuum is commonly denoted as $c = 2.99 \times 108$ meters per second. The phenomenon of light propagation in a vacuum is characterized by a consistent velocity, which is widely acknowledged as an immutable universal constant. The speed of light undergoes alterations when it traverses nonvacuum mediums, such as air (experiencing a decrease of 0.03%) or glass (experiencing a decrease of 30.0%) (Schmidt & Cuevas, 1999).

In the context of various applications, it is common to describe light in relation to its direction and magnitude. In a vacuum, the propagation of light occurs along a linear trajectory at a constant velocity, facilitating the transfer of energy between spatial locations. Two fundamental characteristics of the interaction between light and a medium are as follows: (Sulaiman et al., 2015).

- Light can undergo a change in direction when it transitions from one medium to another, a phenomenon known as refraction.
- Reflection occurs when an object's trajectory is altered upon encountering a surface, resulting in the bouncing back of the object.

The discipline concerned with the identification and quantification of light energy is referred to as radiometry. A standardized system is employed for the characterization of radiant energy. Table 4.1 provides a comprehensive definition of the standard terminology employed throughout the duration of this academic course (Perni et al., 2009).

Table 4.1. Definition and Units for Radiometric (Source: Bahaa E. A. Saleh, Creative Commons License)

Term	Definition	Symbol	Units
Quantity	Radiant energy	Q	Joule (J)
Flux	Rate of radiant energy	Φ	Watt (W); Joule/second (J/s)
Flux density (irradiance)	The flux per unit area	E	Watts per square meter (W/m^2)
Intensity	Flux per solid angle	I	Watts per steradian (W/sr)
Radiance	Flux per unit area per unit solid angle	L	Watts per square meter per steradian ($W/m^2 \cdot sr$)
Spectral radiance	Radiance per unit wavelength	Lλ	Watts per square meter per steradian per nanometer $\frac{W}{m^2 \cdot sr \cdot \Delta\lambda}$

4.3.1. Dual Nature of Light

Researchers construct mathematical models of physical phenomena in order to enhance their comprehension and prognostication of said phenomena. In a similar manner, this phenomenon can be observed with light energy. The observation of light's effects facilitates the development of models. Researchers have made observations indicating that light energy exhibits dual behavior, manifesting as both a wave when traversing through space and as discrete particles possessing quantized energy that can be absorbed and emitted. In the context of studying and utilizing light, both models provide valuable assistance (Sfard, 1991).

Remember

Light exhibits both wave-like and particle-like properties, known as the wave-particle duality. This duality is a fundamental concept in quantum mechanics and helps explain various phenomena like the photoelectric effect.

Photons are employed as a model to represent the particle-like characteristics of light. A photon possesses neither mass nor charge. The phenomenon under consideration functions as a medium for the transmission of electromagnetic energy and exhibits interactions with various discrete entities (de Oliveira Castro et al., 1985). The phenomenon of a light beam can be conceptualized as a collection of individual photons, with each photon possessing a distinct energy value that is contingent upon the wavelength of the light. The energy associated with a specific photon can be determined through the following calculation:

$$\text{Photon energy } (E) = hc/\lambda \tag{4.1}$$

where E is in joules; h = Planck's constant = 6.625 10^{-34} J·s; c = Speed of light = 2.998 × 10^8 m/s; λ = light's wavelength in meters.

The emission of electrons is observed when certain metal surfaces are exposed to ultraviolet light. The aforementioned phenomenon is illustrated in Figure 4.1. The outcomes of the photoelectric effect were inconsistent with the initial predictions derived from wave theory. There were two concerns (Rebelo et al., 2007):

- The increase in radiation intensity, characterized by larger-amplitude waves, did not result in a corresponding increase in the energy of emitted electrons.
- The energy of the emitted electron depended on the light's wavelength.

In the experimental setup depicted in Figure 4.1, incident light impinges upon a metallic plate, thereby initiating the photoelectric effect. The liberation of electrons occurs instantaneously. The measurement of electric current in the external circuit allows for the determination of the number of electrons produced in response to a specific light signal (Taggart & Robey, 1981).

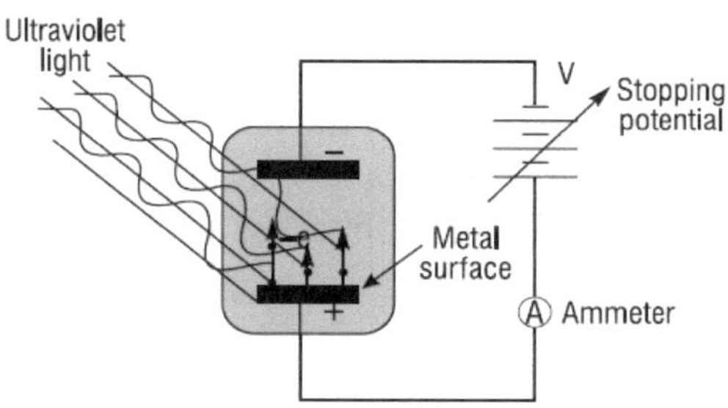

Figure 4.1. Photoelectric effect experiment (Source: Bahaa E. A. Saleh, Creative Commons License).

In the scenario where light is considered as a continuous wave, it is plausible that it would impinge upon the metal surface and engage in interactions with the electrons, thereby providing them with the requisite energy to disengage from the surface, particularly at diminished levels of light intensity. However, it is important to note that this process may exhibit substantial temporal delays. Nevertheless, the emission of electrons was promptly triggered by the presence of low-intensity light with high frequencies corresponding to short wavelengths. Consequently, the photons, acting as discrete entities, displaced the electrons from the metal surface through the phenomenon of light-matter interaction (Maitre et al., 2017).

The electron's ability to escape from the metal is contingent upon surpassing a minimum energy threshold. Electrons are not emitted by photons with frequencies below a specific threshold, regardless of the intensity of the light. Regardless of the intensity, photons with frequencies above the threshold are capable of ejecting electrons. The energy of the emitted electrons can be determined through calculation (Raiteri et al., 2021):

$$E_{e-} = hc/\lambda - p \quad (4.2)$$

where, p = characteristic escape energy for the metal; E_{e-} = the kinetic energy of an escaping electron; hc/λ = the energy of the photon of wavelength λ.

4.3.2. Wave Model

The particle-like model of light elucidates macroscopic phenomena, such as the transmission of light through lenses or its reflection off mirrors. Nevertheless, it is imperative to employ a wavelike model to elucidate intricate phenomena such as interference and diffraction, manifesting when light traverses minute apertures or encounters sharp boundaries. The phenomenon of light or electromagnetic energy transmission in space can be explained by the concept of wave propagation. The propagation of energy through a wave occurs without the transfer of mass, and this process is characterized by a speed that remains constant regardless of the wave's intensity or wavelength. The wave-like characteristics of light form the fundamental principles of physical optics, encompassing the study of light's interaction with various media. A rigorous description of many of these processes necessitates the utilization of calculus and quantum theory. In order to facilitate the practical implementation of photonics technology, it is deemed satisfactory to present the derived equations and models that can be effectively employed by photonics technicians in real-world applications (Ris et al., 1999).

Tip

The wave model of light describes its propagation as an electromagnetic wave, which can undergo various phenomena such as reflection, refraction, diffraction, and interference.

4.3.2.1. Properties of Light Waves

One must have a firm grasp on the basics of wave motion in order to make sense of light waves. Crests and troughs, or high and low points, are the distinguishing features of water waves as they move across the water's surface. Parallel patterns of crests and troughs can be seen along the beach as waves approach the coast. An electromagnetic wave is made up of a magnetic and an electric field that are both pulsating and changing amplitude at regular intervals. Similar to how water particles oscillate vertically within a horizontal wave, the fields' orientations are transverse to the wave transmission's direction (Bekshaev, 2000).

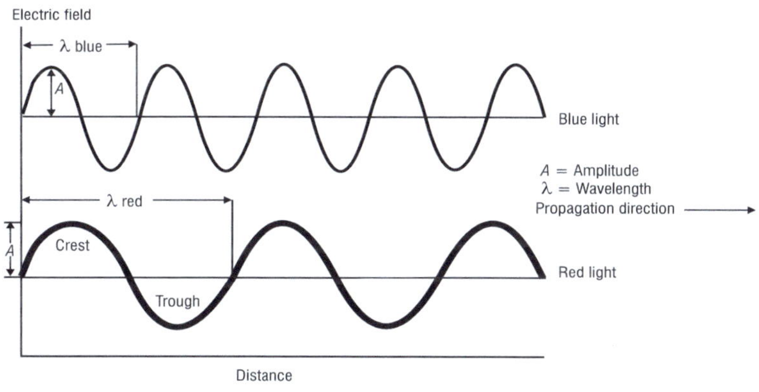

Figure 4.2. 1D view of electromagnetic wave (Source: Bahaa E. A. Saleh, Creative Commons License).

A wave's amplitude (A) refers to its maximum wave displacement value. The cycle commences at a value of zero and subsequently recurs after a specified interval. The term used to refer to this particular measure of distance is known as the wavelength (λ). Various wavelengths of light can be observed, as exemplified by the blue and red light depicted in Figure 4.2. The reciprocal of the wavelength ($1/\lambda$) corresponds to the wave number (v), denoted in units of cm^{-1}. The propagation of the wave occurs at a velocity denoted as the wave speed (v). The velocity of a wave in a vacuum is denoted as c, and it is lower than c when the wave propagates through a medium. The duration required to finish a single cycle is referred to as the cycle time or period (τ) and can be determined through the utilization of Eq. (4.3) (Liu, 1997).

$$\tau = \lambda/v \quad (4.3)$$

The frequency (f) is another significant parameter used to characterize a wave. The quantity is quantified as the frequency of wave cycles traversing a specific location within a time span of one second. The unit of measurement for frequency is cycles per second, which is commonly referred to as hertz (Hz). It is evident that the frequency and the period exhibit a reciprocal relationship. Eq. (4.4) can be utilized to calculate the frequency when the wave speed and wavelength are known.

$$f = 1/\tau = v/\lambda \quad (4.4)$$

Alternative to sinusoidal shapes, waves can exhibit various forms. Nevertheless, it is crucial to bear in mind that light waves are characterized by transverse magnetic and electric fields that undergo spatial and temporal alterations (Peng et al., 2002).

4.3.2.2. Concept of Light Waves

Light waves exhibit a complex nature. A single dimension does not characterize the waves in question but rather consists of magnetic and electric fields that are mutually perpendicular to each other (Figure 4.3).

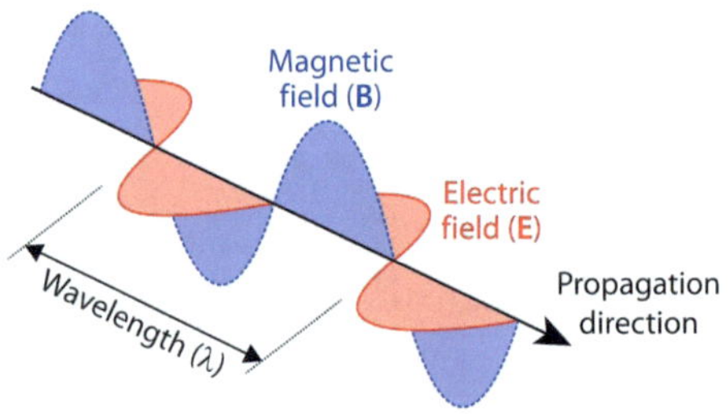

Figure 4.3. Electric and magnetic fields in a light wave (Source: Geert J. J. Verhoeven, Creative Commons License).

All types of wave motion share six characteristics with electromagnetic waves. These properties are discussed below (Mešić et al., 2016):

4.3.3. Polarization

Polarization is a phenomenon that occurs due to the alignment of the electric field vector in relation to the direction of propagation of light. The designation of a light wave as a transverse wave arises from the fact that its electric field oscillates in a direction perpendicular to its propagation direction. Consequently, it possesses the property of being polarizable. In contrast, a sound wave exhibits vibrations in a reciprocal manner along its direction of propagation, thereby rendering it non-polarizable (Bhandari, 1997).

Unpolarized light is characterized by the presence of vibrations occurring in numerous directions, lacking any discernible preferred orientation. Please refer to Figure 4.4(a). Several light sources, such as incandescent bulbs, arc lamps, and the sun, emit light that is unpolarized. Figure 4.4(b) depicts the presentation of vertically polarized light, while Figure 4.4(c) illustrates the manifestation of horizontally polarized light. Each of these instances represents an illustration of linearly polarized light. In Figure 4.4(d), the representation depicts the presence of linearly polarized light at an angle θ relative to the vertical axis (Monteiro et al., 2017).

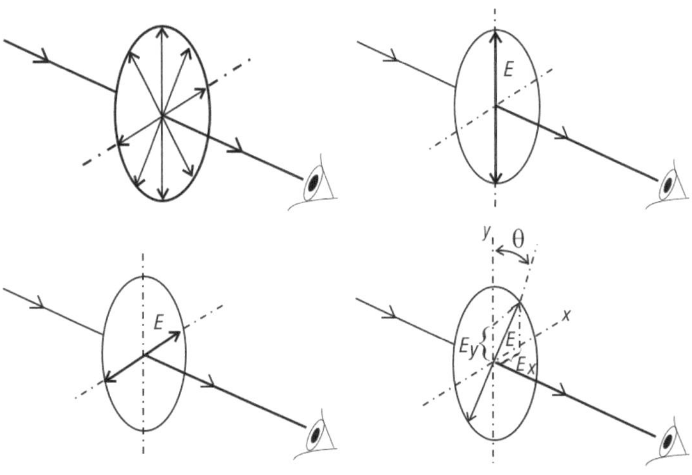

Figure 4.4. Unpolarized and linearly polarized light. The light that is (a) randomly pulsating, (b) unpolarized, (c) linearly polarized down the horizontal, and (d) linearly polarized at an angle to the vertical (Source: Malvin Carl Teich, Creative Commons License).

In certain instances, it is observed that E_x and E_y exhibit a lack of synchrony in their phase, characterized by their failure to attain maximum and minimum values simultaneously. Consequently, the electric field (E-field) does not persist in a constant, linear orientation. In contrast, the maximum amplitudes of the two components do not coincide temporally, resulting in the manifestation of elliptically polarized light. This phenomenon implies that the polarization orientations of light vary as time progresses. Circular polarization is a specific instance of elliptical polarization, characterized by the equality of E_x and E_y, with a phase difference of 90° between them (Li et al., 2018).

Certain materials exhibit the property of selectively transmitting specific polarizations. These devices are commonly referred to as polarizers or analyzers and possess a wide range of applications. When exposed to randomly polarized light, a polarizer selectively transmits light of a specific polarization while absorbing or reflecting light of other polarizations. Polarizing sunglasses serve as a prevalent illustration of the application of polarization in our everyday existence. The lenses allow the transmission of light with electric field vibrations that are perpendicular to specific molecular alignments while simultaneously absorbing light with electric field vibrations that are parallel to these molecular alignments. The primary constituent of light that is reflected from a surface predominantly exhibits horizontal polarization, aligning parallel to the surface (Perrin, 1942).

Consequently, the phenomenon of polarization in sunglasses, wherein the transmission axis is oriented vertically, serves to selectively block horizontally polarized light, thereby diminishing the intensity of glare. Nevertheless, when contemplating a sunbather in a lateral position who is donning sunglasses of this nature, the customary vertical polarization will assume a 90° orientation, aligning parallel to the surface. Consequently, the horizontally polarized light that is reflected from the water or land will be permitted to pass through (Wu et al., 2015).

The calculation of the light intensity that traverses a linear polarizer can be determined by:

$$I(\theta) = \mathrm{I0}\ \cos 2\ (\theta) \tag{4.5}$$

The variable I(θ) represents the light intensity that is transmitted through the polarizer. The variable I0 represents the intensity of the incident light.

The measurement of the angle between the electric field (E-field) and the transmission axis is denoted as θ (Gao et al., 2019).

4.3.4. Huygens' Principle

During the 17th century, Christian Huygens postulated a principle that enables the anticipation of the future position of a given wavefront, provided the current location is known (Figure 4.5). The underlying assumption of his principle is that every point on a wavefront can be regarded as a point source, generating secondary spherical wavelets. Following a certain duration, the updated location of the wavefront will correspond to the surface that is tangential to the secondary wavelets (Pao & Varatharajulu, 1976).

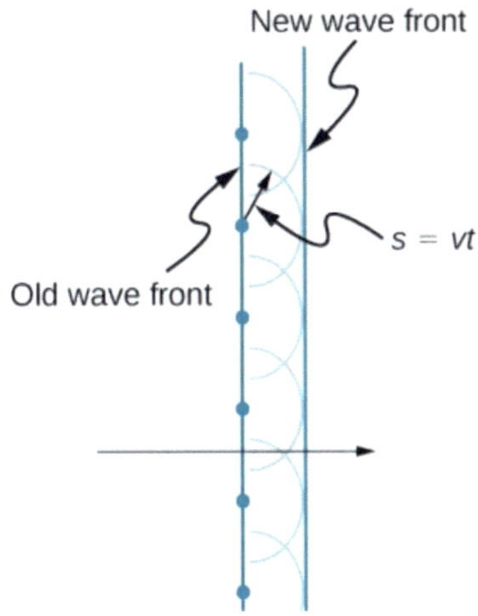

Figure 4.5. Huygens' principle for creating new wavefronts (Source: Rice University, Creative Commons License).

4.3.5. Superposition

In various wave phenomena, such as electromagnetic waves, it is possible for multiple waves to coexist and propagate through the same spatial region simultaneously, exhibiting independent behavior. This implies that the electric field at any given point in space can be determined by calculating the vector sum of the electric fields generated by the individual waves at that particular point. The concept being referred to is the superposition principle. The superposition principle is satisfied by both the magnetic and electric fields of an electromagnetic wave. Therefore, when multiple waves are present, the field at a specific point can be determined by adding together the individual wave vectors (Pi & Orriols, 1985). The phenomenon that occurs when multiple waves are combined and interact with each other is referred to as interference. Consider a scenario where two waves, denoted as y_1 and y_2, possess wavelengths and phases that are almost identical, meaning that their maxima occur at nearly the same time and location. The superposition of these waves yields a resultant wave ($y_1 + y_2$) with an amplitude nearly twice that of the individual waves. Please refer to Figure 4.6a. The phenomenon being referred to is commonly known as constructive interference. When the maximum value of one wave is in close proximity to the minimum value of another wave, the combined amplitude of the resultant wave ($y_1 + y_2$) is significantly reduced, as depicted in Figure 4.6b. The phenomenon being referred to is commonly known as destructive interference (Yue et al., 2017).

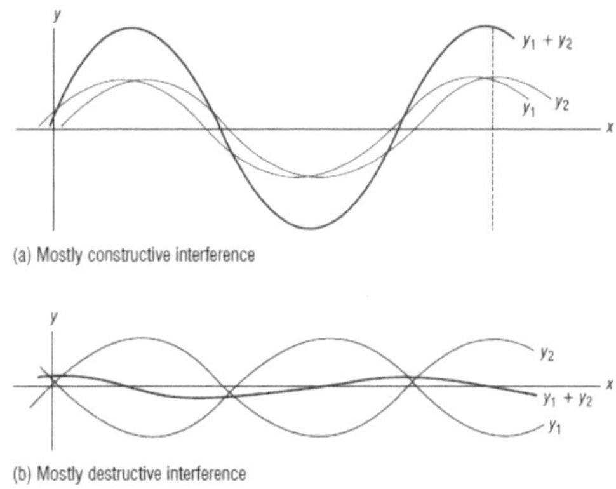

Figure 4.6. Using the principle of superposition to add individual waves (Source: Malvin Carl Teich, Creative Commons License).

4.3.6. Reflection

The alteration in the trajectory of a light ray upon reflection from a surface, such as a mirror, is solely determined by the angle of incidence. According to the principle of reflection, the incident angle upon a reflective plane surface is equivalent to the angle of reflection (Bartell et al., 1968).

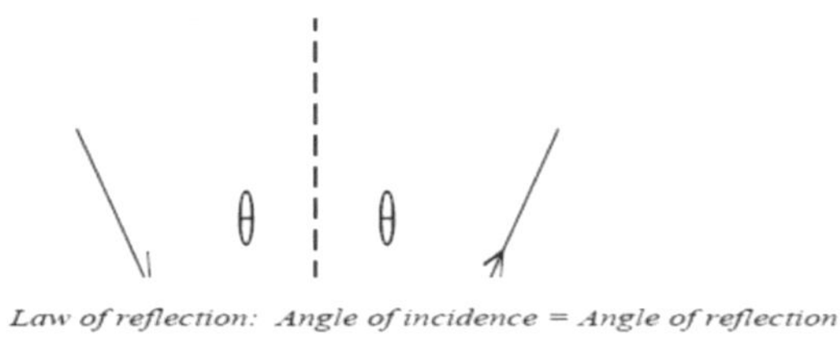

Figure 4.7. Law of reflection: Angle of incidence = Angle of Reflection (Source: Malvin Carl Teich, Creative Commons License).

4.3.7. Refraction

When light transitions from one medium to another, it undergoes a change in direction, commonly referred to as bending, at the interface. This phenomenon occurs due to the disparity in wave speed between the two media. The quantity denoting the ratio of the speed difference is commonly referred to as the index of refraction (n). The relationship between the indices of refraction and the direction of the two rays of light in the two media is mathematically represented by Snell's law (Valanju et al., 2002).

$$\frac{n2}{n1} = \frac{\pm\ q}{\pm\ j} \tag{4.6}$$

where, n_1,n_2 = indices of refraction for the two media; θ = angle of incidence; and φ = angle of refraction.

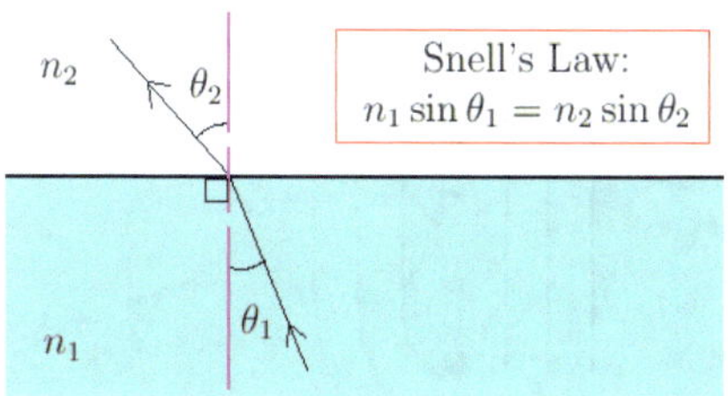

Figure 4.7. Snell's law and Refraction (Source: UBC Math, Creative Commons License).

4.3.8. Diffraction

The wave model's accuracy was substantiated by the elucidation of observed phenomena, such as diffraction and interference, providing conclusive evidence. When light encounters an obstacle, the resulting shadow does not exhibit the characteristics that would be expected according to geometrical ray theory. Instead, the light is diffracted to some extent into the dark area located behind the obstacle, resulting in a shadow with an indistinct boundary. Diffraction, characterized by its tendency to disperse when passing through narrow apertures or encountering sharp edges, can be elucidated by considering the wave-like nature of light. Diffraction can be accurately described and predicted by Huygens' principle. Figure 4.8 depicts the scenario where a wave is approaching a barrier from the left. The barrier is equipped with a narrow opening. Each location on the incident wavefront that reaches the aperture can be considered the location of a propagating spherical wavelet. When the size of an aperture is significantly smaller than the wavelength, the aperture behaves as a point source, giving rise to spherical waves. As the width of the slit, denoted as d, increases, the diffracted wave increasingly resembles the incident plane wave, with the exception of the edges located at the shadow region (Muiño, 2000).

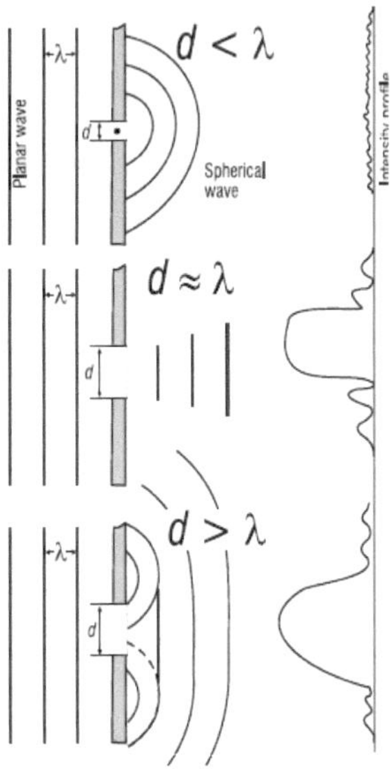

Figure 4.8. Diffraction through slits of differing size (Source: Bahaa E. A. Saleh, Creative Commons License).

4.3.9. Interference

The initial conclusive exhibition of the wave-like characteristics of light was presented through the classical two-slit experiment conducted by Thomas Young in the year 1801. The dimensions of the two slits are significantly smaller in comparison to the distance between them. Hence, every aperture generates diffracted spherical waves that superimpose as they propagate into the region situated to the right of the obstruction. When there is an overlap between the waves, they exhibit interference, resulting in the creation of areas where the waves reinforce each other. These are observed on the display as areas exhibiting the highest levels of intensity. A region of minimum intensity can be observed between adjacent maxima. Please refer to Figure 4.9. The observed pattern displayed on the screen provides a visual representation of locations where constructive interference transpires, denoted as maxima (labeled as B), as well as locations where destructive interference transpires, denoted as minima (labeled as D). The experimental arrangement depicted in Figure 4.9 may be employed in practical applications for the purpose of measuring the wavelength of light (Kolobov & Sokolov, 1986).

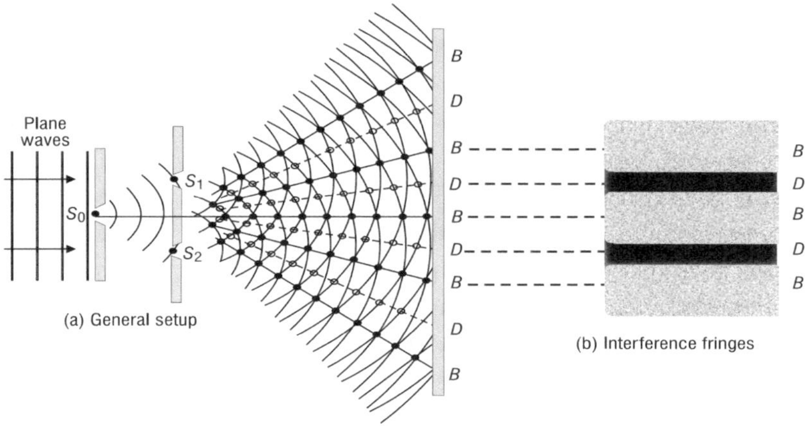

Figure 1-10 *Classic double-slit experiment*

Figure 4.9. Classic double slit experiment (Source: Skulls in the Star, Creative Commons License)

4.4. THE ELECTROMAGNETIC SPECTRUM

All forms of electromagnetic radiation exhibit similar wave-like characteristics, with the only distinguishing factor being their respective wavelengths. Electromagnetic waves span a wide range of wavelengths, extending from those of considerable length, such as the radiation emitted by electric power lines operating at a frequency of 60 Hz, to those of extremely short length, such as gamma-ray radiation. The comprehensive collection of electromagnetic waves is called the electromagnetic spectrum. The spectrum depicted in Figure 4.10 is partitioned into distinct practical applications based on predetermined frequency ranges, which are established through conventional means by the sources and detection devices (Zwinkels, 2015).

The region spanning from infrared to ultraviolet is of utmost significance in photonics. Nevertheless, it is important to acknowledge that every regime possesses a certain degree of usefulness. The rotation of generators and power lines results in the generation of waves with low frequencies. The wavelengths under consideration range from 10^5 to 10^8 meters. Heinrich Hertz generated electromagnetic waves within a highly practical range of wavelengths from 0.3 to 105 meters. Radio broadcasting bands are typically located within the lower range of wavelengths. The microwave spectrum encompasses wavelengths ranging from 0.01 to 0.3 meters, which are utilized for satellite communication purposes. The discovery of the infrared region, spanning from 1 μm to 30 μm, is attributed to Sir William Herschel, who initially detected it in the year 1800. The geographical area under consideration is partitioned into five distinct regions, namely: the very near-infrared region (1–3 μm), the near-infrared region (3–5 μm), the mid-infrared region (5–6 μm), the far infrared region (6–15 μm), and the very long infrared region (15–30 μm). Similar to the auditory system's inability to perceive sounds beyond specific frequencies, the visual system of humans is limited in its ability to perceive light outside a narrow spectrum of wavelengths, specifically ranging from 0.76 to 0.49 μm. The ultraviolet region, which was first observed by Johann Ritter, is a region characterized by higher energy levels. The phenomenon under consideration elicits numerous chemical reactions and is responsible for the ionization of the upper atmosphere, thereby giving rise to the ionosphere (Gertsenshtein, 1962).

Did You Know?

The electromagnetic spectrum encompasses a wide range of electromagnetic waves, from radio waves with long wavelengths to gamma rays with extremely short wavelengths. Each portion of the spectrum has unique properties and applications, making it essential for various technologies and scientific studies.

The X-ray regime was discovered by Wilhelm Röntgen in the year 1895. The wavelength of this phenomenon spans from 10^{-8} to 10^{-11} meters. Due to its considerable energy level, this particular form of radiation possesses the ability to permeate organic tissue and generate visual representations of materials with greater density, such as bones. Gamma rays are characterized by having the shortest wavelength, measuring less than 10^{-13} meters. Solar radiation, linear and particle beam accelerators, and nuclear processes are sources that emit high-energy particles, which display characteristics similar to particles (Diddams et al., 2020).

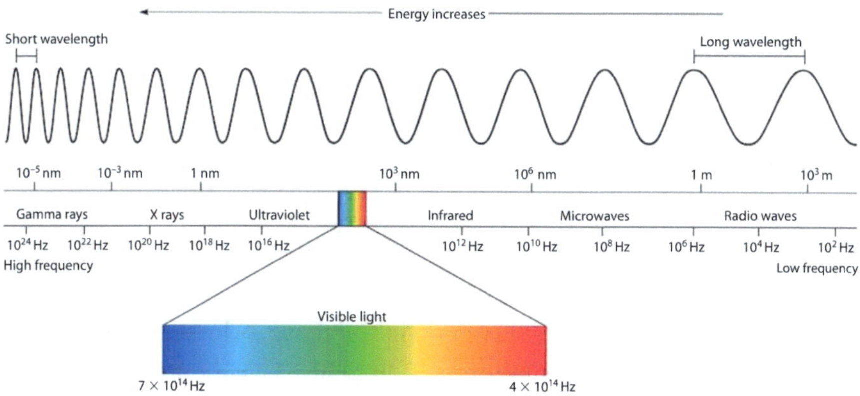

Figure 4.10. Electromagnetic spectrum (Source: Mini Physics, Creative Commons License).

White light is composed of a combination of various colors of light (Figure 4.11). Each color possesses a distinct wavelength and, upon traversing a transparent medium, exhibits varying degrees of refraction. Hence, it is observed that a prism has the capability to disperse white light into its constituent colors (Herter, 1985).

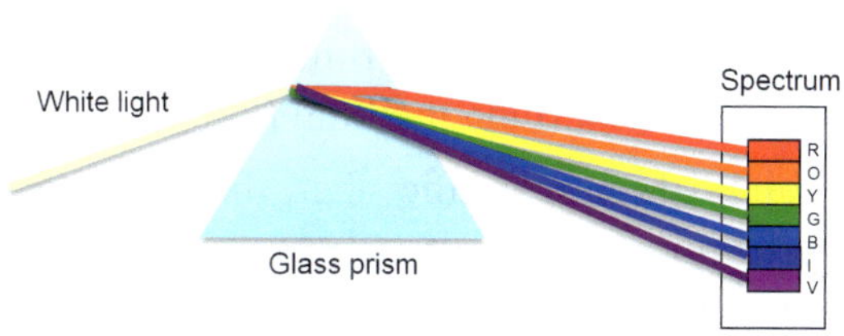

Figure 4.11. Separation of light into component colors (Source: Toppr, Creative Commons License).

Table 4.2. Visible Spectrum Wavelengths (Source: Bahaa E. A. Saleh, Creative Commons License)

Color	**Wavelength Band (μm)**	**Representative wavelength (μm)**
Extreme violet	0.39–0.41	0.40
Violet	0.39–0.45	0.43
Dark blue	0.45–0.48	0.47
Light blue	0.48–0.50	0.49

Green	0.50–0.55	0.53
Yellow-green	0.55–0.57	0.56
Yellow	0.57–0.58	0.58
Orange	0.58–0.62	0.60
Red	0.62–0.70	0.64
Deep red	0.70–0.76	0.72

The categorization of colors observed in visible light is based on their respective wavelengths. The wavelengths of these colors are provided in Table 4.2. The term "spectrum of light" refers to a visual representation that depicts the distinct constituents of light arranged in a sequential order based on their respective wavelengths (Goulielmakis et al., 2004).

4.5. SPECTRA OF LIGHT SOURCES

There exists a multitude of diverse sources that emit electromagnetic radiation. Typically, sources are classified into two distinct classes: natural sources and man-made sources. In the field of photonics, two significant spectra hold importance: the emission spectrum and the absorption spectrum. The emission spectrum is derived from the electromagnetic radiation emitted by a given source. The absorption spectrum is derived from analyzing the light that has traversed an absorbing medium (Seliger, 1960).

It is a well-established scientific fact that all materials can emit electromagnetic radiation when their temperatures exceed absolute zero. Each atom and molecule possess a distinct and unique arrangement of spectral lines. The observation of line spectra during the early stages of scientific exploration yielded valuable insights into the fundamental properties of atoms. The discoveries made by Planck and Einstein played a pivotal role in the advancement of contemporary quantum theory.

The fingerprints of these objects can exhibit various combinations of spectral lines, bands, and continuums. The phenomenon of atoms undergoing state transitions results in the emission of both visible and ultraviolet radiation. Infrared radiation is generated as a result of the transition of molecules between different vibrational and rotational states. Dense materials exhibit a multitude of energy states, resulting in emission and absorption bands that span a wide range for both solids and liquids. In the case of gas with lower density, the spectral bands exhibit a significantly reduced width (Fotios et al., 2005).

In order to view a band spectrum, a beam of light is directed through a narrow aperture. The optical phenomenon of refraction or diffraction is observed when the image of the slit is subjected to a prism or a grating, respectively, depending on the specific wavelengths present in the incident light. The phenomenon is captured through the use of either film or a spectrograph. The lines connect with the fundamental atomic structure and the distinctive alterations in energy levels. Spectroscopy is an academic discipline that involves analyzing line spectra and identifying various constituents within materials (Jang et al., 1999).

When atoms are in proximity to one another, their electrons engage in interactions, resulting in the division of energy levels. Within a solid material, there exist numerous energy levels that facilitate the emission or absorption of a continuous spectrum of frequencies. Materials that possess high temperature and density have the ability to emit spectra that are continuous in nature, consisting of various bands of frequencies (Nikolaev et al., 2005).

4.5.1. Atomic Structure

The fundamental constituents of all matter are atoms. The atom is the fundamental unit that preserves the distinctive properties of a chemical element. The atomic structure comprises a positively charged nucleus encompassed by negatively charged electrons organized in discrete energy levels denoted as K through O, as depicted in Figure 4.12. The symbol K(2) denotes the state in which the K-shell achieves electron saturation with a total of two electrons (Sleator et al., 1992).

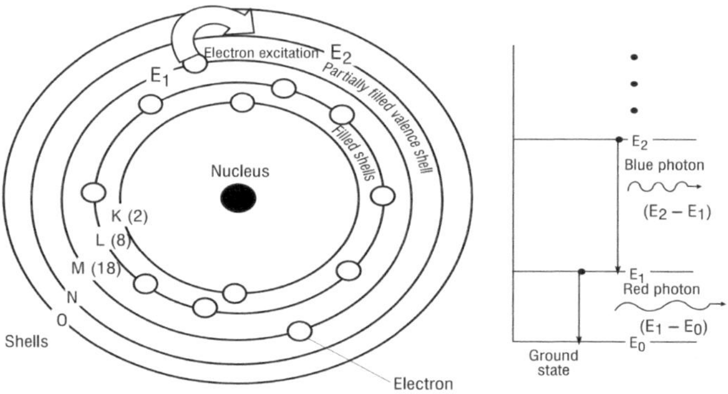

Figure 4.12. Atomic Model (Source: Malvin Carl Teich, Creative Commons License).

The energy of an atom is represented by a model that incorporates the behavior of its electrons. The atom is considered to be at its lowest energy level when all of its electrons are in the ground state. When an atom assimilates energy, it has the capacity to "excite" its electrons, causing them to transition to higher-energy orbitals. When electrons transition between different energy levels, discrete quantities of energy, known as quanta, are either absorbed or emitted. This is the process by which an atom can undergo absorption or emission of light. The energy quanta of light are contingent upon the electronic configuration of the atom (Rasel et al., 1995).

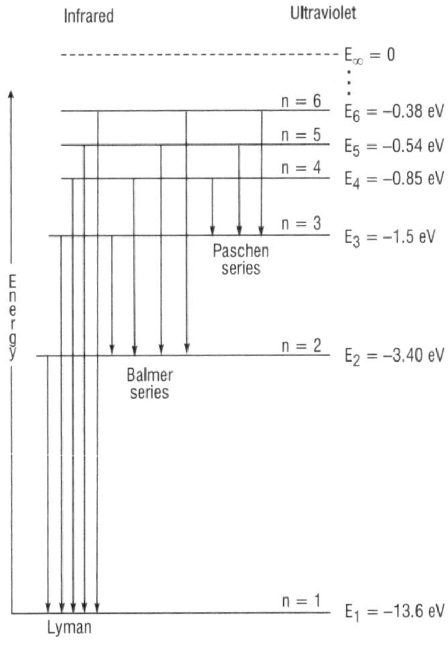

Figure 4.13. Energy-level diagram for a hydrogen atom (Source: Toppr, Creative Commons License).

The atomic energy-level diagram illustrates the distinct electron energy states that are accessible within a specific atom. Figure 4.13 illustrates an energy-level diagram of hydrogen. Hydrogen possesses a solitary electron capable of occupying solely one of the energy levels depicted at any given moment. The ground state denoted as E1, represents the lowest energy level. In order for the electron to transition to a higher energy level, an input of energy is required for the atom. It should be noted that the energy levels span a spectrum, starting from -13.6 electron volts (eV) for the lowest energy level (n = 1) and reaching 0 eV for the utmost energy level (n = ∞), which signifies the electron's liberation from the atom (Oberthaler et al., 1996).

Consider a hydrogen atom that is in an excited state, specifically the n = 3 energy level. The ground state of an atom can be achieved through the emission of a photon during a transition. The energy of a photon is equivalent to the alteration in energy experienced by the atom, as denoted by Eq. (4.7).

$$Ephoton = E\ E3 - 1 = -1.51\ \text{eV} - (-13.6\ \text{eV}) = 12.09\ \text{eV} \tag{4.7}$$

The atom can also absorb photons with energies that precisely correspond to disparities between electron energy levels. As an illustration, it is possible for a hydrogen atom in its lowest energy state, known as the ground state, to undergo absorption of a photon with an energy of 12.09 electron volts (eV). The electron within the atom undergoes a transition from energy level E1 to energy level E3 (Altshuler et al., 1966).

4.5.2. Molecular Structure

Molecules present in gaseous or liquid states have the capability to absorb electromagnetic radiation as well. In order for a photon to interact with a molecule, its energy must correspond to a specific and quantized rotational energy level of the molecule. Absorption in solids is a multifaceted phenomenon, typically arising from alterations in vibrational energy states.

4.6. BLACKBODY RADIATION

To initiate the process of comprehending blackbody radiation, it is imperative to elucidate the interconnections among temperature, wavelength, and energy discharged by an ideal thermal radiator, commonly referred to as a blackbody. Based on empirical observations, it is understood that objects with varying temperatures emit radiation characterized by distinct wavelengths or colors. As an illustration, the metallic conductors within a heating device undergo a transformation wherein they emit a red glow subsequent to being subjected to thermal energy.

Blackbody radiation refers to the theoretical upper limit of radiation that can be emitted by an object due to its temperature-induced thermal self-radiation. The radiation in question exhibits a distribution of peak energy that can be observed within the infrared, visible, or ultraviolet regions of the electromagnetic spectrum. As the temperature of the emitter increases, there is a corresponding increase in the amount of energy emitted and a decrease in the wavelength of the emitted radiation. An item at standard ambient temperature exhibits its maximum radiation in the infrared spectrum, whereas the sun's maximum radiation occurs within the visible spectrum (Einstein, 1905).

The equations utilized for computing radiation about temperature are formulated based on the Kelvin temperature scale. Please ensure that all calculations are conducted using the Kelvin scale. The equations 4.8 and 4.9 present the mathematical expressions for converting between various temperature scales.

Fahrenheit (F) to Celsius (C) $°C = 5/9\,(°F - 32)$ (4.8)

Celsius (C) to Kelvin (K) $K = °C + 273.15$ (4.9)

A waveband refers to a specific electromagnetic spectrum segment delineated by predetermined upper and lower wavelengths. The total energy emitted by a blackbody within a specific range of wavelengths is the cumulative sum of all energies emitted at individual wavelengths falling within that range. The rate at which energy is emitted through radiation is referred to as the power of radiation. Additionally, the total power radiated by a blackbody can be determined by incorporating the influence of all emitted wavelengths. The power emitted per unit surface area of a blackbody at temperature T can be determined using Eq. (4.10), which is known as the Stefan-Boltzmann law.

$$Ws = \sigma s\, T^4 \text{ watts/m}^2 \quad (4.10)$$

where $\sigma_s = 5.67 \times 10^{-8}$ watts/m^2·K^4

T = Temperature (K)

The emitted radiant flux density, denoted as Ws, refers to the power per unit area. A graybody refers to an object that does not exhibit the characteristics of a perfect "blackbody" but instead emits radiation at a fraction of the theoretical maximum emitted by a blackbody. The radiant flux density emitted by a blackbody is diminished by a factor known as the emissivity. The emissivity (ε) is contingent upon the emitting material and is a value that is lower than 1. Therefore, the radiant flux density emitted by a graybody can be mathematically represented by Eq. (4.11) (Cornelius & Dowling, 1999).

$$Ws = \varepsilon\, \sigma s\, T^4 \text{ watts/m}^2 \quad (4.11)$$

4.6.1. Spectral Distribution

The emission of radiation from a blackbody is distributed across a range of wavelengths. The term "spectral flux density" refers to the quantity W_λ.

The power radiated per unit area of a surface for wavelengths within the waveband Δλ is denoted as Wλ Δλ. Max Planck formulated an equation in the year 1900 that exhibits a remarkable degree of accuracy in fitting experimental observations of Wλ. The radiation formula proposed by Planck is represented by Eq. (4.12).

$$W\lambda = \frac{C_1}{\lambda^5} \frac{1}{e^{\frac{C_2}{\lambda T}} - 1} \tag{4.12}$$

where, λ = wavelength (m); T = blackbody temperature (K); $C_1 = 2\pi c^2 h = 3.75 \times 10^{-16}$ W·m²; $C_2 = hc/k = 1.44 \times 10^{-4}$ m·K; c = speed of light = 3.00×10^8 m/s; h = Planck's constant = 6.626×10^{-34} J·s; k = Boltzmann's constant = 1.38×10^{-23} J/K.

Figure 4.14 illustrates the spectral flux density emitted by a blackbody, as determined by Planck's formula, across a range of five distinct temperatures. The wavelengths are represented on the graph using microns as the unit of measurement. It should be noted that the axes exhibit a logarithmic scale (Strömgren, 1976).

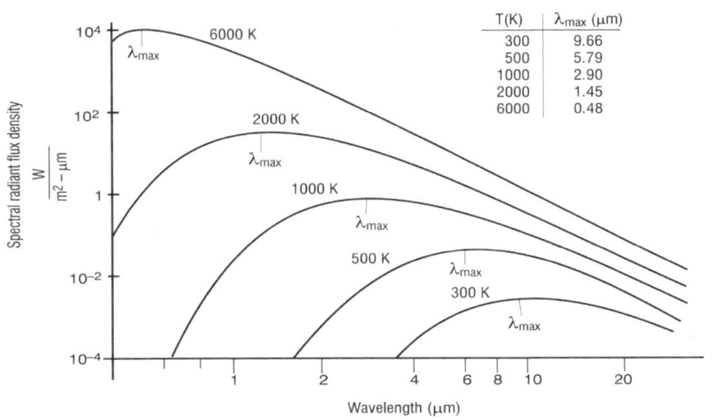

Figure 4.14. The present study investigates the spectral radiant blackbody flux density variations at different temperatures (Source: V. Roberts, Creative Commons License)

4.6.2. Wien's Displacement Law

Each blackbody temperature is associated with a spectral distribution that exhibits a maximum emission wavelength, also known as the peak wavelength. The maximum wavelength is correlated with the temperature of the blackbody. The relationship can be described by Wien's displacement law:

$$\lambda_{max} T = 2.898 \times 10^{-3}\ m{\cdot}K \qquad (4.13)$$

According to Wien's displacement law, the peak wavelength of a blackbody's emission spectrum is inversely proportional to its temperature, indicating that as the temperature of the blackbody increases, the peak wavelength decreases.

4.7. INTERACTIONS OF LIGHT WITH MATTER

When light propagates through a medium, it undergoes interactions with the constituent particles of the medium. The pivotal interactions encompass scattering and absorption.

4.7.1. Absorption

Absorption refers to the process by which energy is transferred from an electromagnetic wave to the atoms within a given medium. The transfer of energy to an atom has the potential to induce the excitation of electrons, causing them to transition to higher energy states. The transfer of energy to a molecule has the potential to induce excitations in the form of vibrational or rotational motions. The excitation of energy states is contingent upon the energy-level structures, which, in turn, are determined by the composition of molecules and atoms present in the medium. Consequently, these factors influence the wavelengths of light capable of inducing such excitation. The observed light spectrum after its transmission through a medium exhibits the absence of specific wavelengths, which can be attributed to their absorption. The term used to describe this phenomenon is an absorption spectrum.

The phenomenon of selective absorption serves as the fundamental principle underlying the manifestation of color in objects. The red color of an apple is attributed to its ability to selectively absorb wavelengths from the visible spectrum while reflecting predominantly red light (Wood, 1902).

4.7.2. Scattering

Scattering refers to the phenomenon in which the trajectory of light is altered due to its interaction with matter.

When the dimensions of the scatterer, such as a molecule, are significantly smaller than the light's wavelength, it is possible for the scatterer to absorb the incident light and promptly reemit it in a divergent trajectory. When the wavelength of the reemitted light is identical to that of the incident light, the phenomenon is referred to as Rayleigh scattering (Figure 4.15) (Starunov & Fabelinskiĭ, 1970). When the wavelength of the reemitted light is increased, it indicates that the molecule remains in an excited state, and this phenomenon is referred to as Raman scattering. Raman scattering involves the emission of secondary photons with longer wavelengths as a result of the molecule's transition back to its ground state.

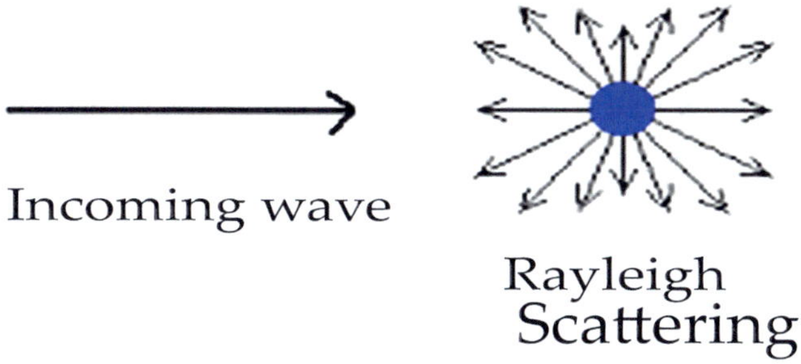

Figure 4.15. Rayleigh scattering (Source: Mike Wills, Creative Commons License).

The scattering of visible light by air molecules, specifically oxygen (O_2) and nitrogen (N_2), follows the Rayleigh scattering phenomenon. This scattering process is more pronounced for shorter wavelengths, such as those in the blue and violet portions of the electromagnetic spectrum (Grünberg et al., 1982).

Activity 4.1

In a photography studio, the photographer aims to capture a portrait with a soft and flattering appearance. Explain how the concepts of diffraction and reflection can be utilized to achieve this effect using appropriate lighting techniques and equipment.

SUMMARY

This chapter introduces the fascinating world of photonics and explores the properties and interactions of light. The chapter begins with an overview of photonics and then delves into the dual nature of light, presenting both its wave and particle characteristics. The properties of light, such as polarization,

Huygens' principle, superposition, reflection, refraction, diffraction, and interference, are discussed in detail, highlighting their essential role in understanding light behavior. The electromagnetic spectrum is introduced, showcasing the various forms of light, from radio waves to gamma rays. The chapter then explores the spectra of light sources, delving into atomic and molecular structures and their role in light emission. Blackbody radiation and its spectral distribution, as well as Wien's displacement law, are explained to understand the thermal radiation of objects. The interactions of light with matter, particularly absorption and scattering, are examined, shedding light on how light interacts with substances it encounters. This chapter provides a comprehensive overview of the fundamental concepts of light waves and sets the stage for deeper exploration in the field of optics and photonics.

REVIEW QUESTIONS

1. What is photonics, and why is it an important area of study in modern science and technology?
2. Describe the dual nature of light and explain how it exhibits both wave-like and particle-like behaviors.
3. Discuss the properties of light, such as polarization, Huygens' principle, and superposition, and their significance in understanding light behavior.
4. How does light interact with surfaces? Explain the processes of reflection, refraction, diffraction, and interference.
5. What is the electromagnetic spectrum, and what are the different types of light waves it encompasses?
6. Describe the spectra of light sources and their relationship with atomic and molecular structures.
7. Explain blackbody radiation and its spectral distribution. How does Wien's displacement law relate to the thermal radiation of objects?

REFERENCES

1. Altshuler, S., Frantz, L. M., & Braunstein, R. (1966). Reflection of atoms from standing light waves. *Physical Review Letters*, *17*(5), 231.
2. Baba, T., & Nakamura, M. (2002). Photonic crystal light deflection devices using the superprism effect. *IEEE Journal of Quantum Electronics*, *38*(7), 909–914.
3. Bartell, L. S., Roskos, R. R., & Thompson, H. B. (1968). Reflection of electrons by standing light waves: experimental study. *Physical Review*, *166*(5), 1494.
4. Bauke, H., Ahrens, S., & Grobe, R. (2014). Electron-spin dynamics in elliptically polarized light waves. *Physical Review A*, *90*(5), 052101.
5. Bekshaev, A. Y. (2000). Manifestation of mechanical properties of light waves in vortex beam optical systems. *Optics and Spectroscopy*, *88*, 904–910.
6. Bhandari, R. (1997). Polarization of light and topological phases. *Physics Reports*, *281*(1), 1–64.
7. Boudiombo, J., Baehr, O., Boudrioua, A., Thevenin, P., Loulergue, J. C., & Bath, A. (1997). Modes of propagating light waves in thin films of boron nitride deposited by plasma enhanced chemical vapor deposition. *Materials Science and Engineering: B*, *46*(1–3), 96–98.
8. Cornelius, C. M., & Dowling, J. P. (1999). Modification of Planck blackbody radiation by photonic band-gap structures. *Physical Review A*, *59*(6), 4736.
9. de Oliveira Castro, G., Martins-Ferreira, H., & Gardino, P. F. (1985). Dual nature of the peaks of light scattered during spreading depression in chick retina. *Anais da Academia Brasileira de Ciências*, *57*(1), 95–103.
10. Diddams, S. A., Vahala, K., & Udem, T. (2020). Optical frequency combs: Coherently uniting the electromagnetic spectrum. *Science*, *369*(6501), 3676.
11. Einstein, A. (1905). Indeed, it seems to me that the observations regarding" blackbody radiation," photoluminescence, production of cathode rays by ultraviolet. *Annalen der Physik*, *17*, 132–148.
12. Fotios, S., Cheal, C., & Boyce, P. R. (2005). Light source spectrum, brightness perception and visual performance in pedestrian environments: A review. *Lighting Research & Technology*, *37*(4), 271–291.

13. Gao, S., Park, C. S., Zhou, C., Lee, S. S., & Choi, D. Y. (2019). Twofold polarization-selective all-dielectric trifoci metalens for linearly polarized visible light. *Advanced Optical Materials*, *7*(21), 1900883.
14. Gertsenshtein, M. E. (1962). Wave resonance of light and gravitional waves. *Sov Phys JETP*, *14*, 84–85.
15. Goulielmakis, E., Uiberacker, M., Kienberger, R., Baltuska, A., Yakovlev, V., Scrinzi, A., ... & Krausz, F. (2004). Direct measurement of light waves. *Science*, *305*(5688), 1267–1269.
16. Grünberg, P., Cottam, M. G., Vach, W., Mayr, C., & Camley, R. E. (1982). Brillouin scattering of light by spin waves in thin ferromagnetic films. *Journal of Applied Physics*, *53*(3), 2078–2083.
17. Hashimoto, K., Yano, T., Shimizu, M., & Nayatani, Y. (2007). New method for specifying color-rendering properties of light sources based on feeling of contrast. *Color Research & Application*, *32*(5), 361–371.
18. Herter Jr, C. A. (1985). The electromagnetic spectrum: A critical natural resource. *Nat. Resources J.*, *25*, 651.
19. Jang, Y. S., Lee, C. H., & Chung, Y. C. (1999). Effects of crosstalk in WDM systems using spectrum-sliced light sources. *IEEE Photonics Technology Letters*, *11*(6), 715–717.
20. Kim, M., Jacob, Z., & Rho, J. (2020). Recent advances in 2D, 3D and higher-order topological photonics. *Light: Science & Applications*, *9*(1), 130.
21. Kolobov, M. I., & Sokolov, I. V. (1986). Interference of light waves possessing sub-Poissonian statistics and the sensitivity of laser gravitational observations. *Zh. Eksp. Teor. Fiz*, *90*, 1889–1899.
22. Li, P., Wu, D., Liu, S., Zhang, Y., Guo, X., Qi, S., ... & Zhao, J. (2018). Three-dimensional modulations on the states of polarization of light fields. *Chinese Physics B*, *27*(11), 114201.
23. Liu, N. H. (1997). Propagation of light waves in Thue-Morse dielectric multilayers. *Physical Review B*, *55*(6), 3543.
24. Maitre, N. L., Key, A. P., Chorna, O. D., Slaughter, J. C., Matusz, P. J., Wallace, M. T., & Murray, M. M. (2017). The dual nature of early-life experience on somatosensory processing in the human infant brain. *Current Biology*, *27*(7), 1048–1054.
25. Meng, Y., Chen, Y., Lu, L., Ding, Y., Cusano, A., Fan, J. A., ... & Ni, X. (2021). Optical meta-waveguides for integrated photonics and beyond. *Light: Science & Applications*, *10*(1), 235.

26. Mešić, V., Hajder, E., Neumann, K., & Erceg, N. (2016). Comparing different approaches to visualizing light waves: An experimental study on teaching wave optics. *Physical Review Physics Education Research, 12*(1), 010135.
27. Monteiro, M., Stari, C., Cabeza, C., & Martí, A. C. (2017). The polarization of light and Malus' law using smartphones. *The Physics Teacher, 55*(5), 264–266.
28. Muiño, P. L. (2000). Introducing the uncertainty principle using diffraction of light waves. *Journal of Chemical Education, 77*(8), 1025.
29. Nikolaev, I. S., Lodahl, P., & Vos, W. L. (2005). Quantitative analysis of directional spontaneous emission spectra from light sources in photonic crystals. *Physical Review A, 71*(5), 053813.
30. Oberthaler, M. K., Abfalterer, R., Bernet, S., Schmiedmayer, J., & Zeilinger, A. (1996). Atom waves in crystals of light. *Physical review letters, 77*(25), 4980.
31. Pao, Y. H., & Varatharajulu, V. (1976). Huygens' principle, radiation conditions, and integral formulas for the scattering of elastic waves. *The Journal of the Acoustical Society of America, 59*(6), 1361–1371.
32. Peng, R. W., Huang, X. Q., Qiu, F., Wang, M., Hu, A., Jiang, S. S., & Mazzer, M. (2002). Symmetry-induced perfect transmission of light waves in quasiperiodic dielectric multilayers. *Applied Physics Letters, 80*(17), 3063–3065.
33. Perni, S., Piccirillo, C., Pratten, J., Prokopovich, P., Chrzanowski, W., Parkin, I. P., & Wilson, M. (2009). The antimicrobial properties of light-activated polymers containing methylene blue and gold nanoparticles. *Biomaterials, 30*(1), 89–93.
34. Perrin, F. (1942). Polarization of light scattered by isotropic opalescent media. *The Journal of Chemical Physics, 10*(7), 415–427.
35. Pi, F., & Orriols, G. (1985). Energy balance in the superposition of light waves with lossless beam splitters. *American Journal of Physics, 53*(7), 667–670.
36. Raiteri, C. M., Villata, M., Carosati, D., Benítez, E., Kurtanidze, S. O., Gupta, A. C., ... & Zhovtan, A. V. (2021). The dual nature of blazar fast variability: Space and ground observations of S5 0716+ 714. *Monthly Notices of the Royal Astronomical Society, 501*(1), 1100–1115.
37. Rasel, E. M., Oberthaler, M. K., Batelaan, H., Schmiedmayer, J., & Zeilinger, A. (1995). Atom wave interferometry with diffraction

gratings of light. *Physical Review Letters*, *75*(14), 2633.

38. Rebelo, L. P. N., Lopes, J. N. C., Esperanca, J. M., Guedes, H. J., Łachwa, J., Najdanovic-Visak, V., & Visak, Z. P. (2007). Accounting for the unique, doubly dual nature of ionic liquids from a molecular thermodynamic and modeling standpoint. *Accounts of Chemical Research*, *40*(11), 1114–1121.
39. Ris, R. C., Holthuijsen, L. H., & Booij, N. (1999). A third-generation wave model for coastal regions: 2. Verification. *Journal of Geophysical Research: Oceans*, *104*(C4), 7667–7681.
40. Rumsey, V. H., Deschamps, G. A., Kales, M. L., Bohnert, J. I., & Booker, H. G. (1951). Techniques for handling elliptically polarized waves with special reference to antennas: introduction. *Proceedings of the IRE*, *39*(5), 533–534.
41. Schmidt, J., & Cuevas, A. (1999). Electronic properties of light-induced recombination centers in boron-doped Czochralski silicon. *Journal of Applied Physics*, *86*(6), 3175–3180.
42. Seliger, H. H. (1960). A photoelectric method for the measurement of spectra of light sources of rapidly varying intensities. *Analytical Biochemistry*, *1*(1), 60–65.
43. Sfard, A. (1991). On the dual nature of mathematical conceptions: Reflections on processes and objects as different sides of the same coin. *Educational studies in mathematics*, *22*(1), 1–36.
44. Sleator, T., Pfau, T., Balykin, V., & Mlynek, J. (1992). Imaging and focusing of an atomic beam with a large period standing light wave. *Applied Physics B*, *54*, 375–379.
45. Soukoulis, C. M. (Ed.). (2012). *Photonic Crystals and Light Localization in the 21st Century*. Springer Science & Business Media. (Vol. 563, pp. 321–325).
46. Starunov, V. S., & Fabelinskiĭ, I. L. (1970). Stimulated mandel'shtam–brillouin scattering and stimulated entropy (temperature) scattering of light. *Soviet Physics Uspekhi*, *12*(4), 463.
47. Strömgren, T. (1976). Length growth of Mytilus edulis (Bivalvia) in relation to phqtoperiod, irradiance, and spectral distribution of light. *Sarsia*, *61*(1), 31–40.
48. Sulaiman, T. A., Abdulmajeed, A. A., Donovan, T. E., Ritter, A. V., Vallittu, P. K., Närhi, T. O., & Lassila, L. V. (2015). Optical properties and light irradiance of monolithic zirconia at variable thicknesses.

Dental Materials, *31*(10), 1180–1187.

49. Taggart, W., & Robey, D. (1981). Minds and managers: On the dual nature of human information processing and management. *Academy of Management Review*, *6*(2), 187–195.
50. Valanju, P. M., Walser, R. M., & Valanju, A. P. (2002). Wave refraction in negative-index media: Always positive and very inhomogeneous. *Physical Review Letters*, *88*(18), 187401.
51. Wood, R. W. (1902). XLIV. A suspected case of the electrical resonance of minute metal particles for light-waves. A new type of absorption. *The London, Edinburgh, and Dublin Philosophical Magazine and Journal of Science*, *3*(16), 396–410.
52. Wu, B., Sung, J. Y., Yan, J. H., Xu, M., Wang, J., Yan, F., ... & Chang, G. K. (2015). Polarization-insensitive remote access unit for radio-over-fiber mobile fronthaul system by reusing polarization orthogonal light waves. *IEEE Photonics Journal*, *8*(1), 1–8.
53. Wurtman, R. J., & Weisel, J. (1969). Environmental lighting and neuroendocrine function: relationship between spectrum of light source and gonadal growth. *Endocrinology*, *85*(6), 1218–1221.
54. Yablonovitch, E. (2001). Photonic crystals: semiconductors of light. *Scientific American*, *285*(6), 46–55.
55. Yue, Z., Qiu, P., Yang, R., Zhang, S., Yuan, K., & Li, Z. (2017). Stress analysis of the interaction of a running crack and blasting waves by caustics method. *Engineering Fracture Mechanics*, *184*, 339–351.
56. Zwinkels, J. (2015). Light, electromagnetic spectrum. *Encyclopedia of Color Science and Technology*, *8071*, 1–8.

CHAPTER 5

Wave Behavior and Interference

Contents

5.1. INTRODUCTION

Waves are intrinsic phenomena that manifest in diverse biological and technological systems, facilitating the transmission of energy and information across a medium. A comprehensive grasp of wave behavior is crucial in disciplines like physics, engineering, as well as communication. This introduction will examine the concepts of interference waves, various types of wave interactions, and the fundamental wave properties, namely amplitude, wavelength, frequency, as well as speed (Shokri-Ghadikolaei & Fischione 2015).

Interference phenomena manifest when multiple waves interact, resulting in the superposition of their respective displacements. The outcome of this interaction may exhibit either constructive or destructive interference, contingent upon the relative phases as well as amplitudes of the waves involved. Constructive interference occurs when waves combine in a manner that results in the reinforcement of each other, leading to the formation of regions characterized by higher amplitude. On the contrary, destructive interference occurs when the waves undergo cancellation, resulting in the emergence of areas characterized by reduced amplitude (Donati et al., 1973).

Interference waves are a prevalent phenomenon observed in a wide range of both organic and engineered systems. In natural settings, one can observe instances such as the aesthetically pleasing formations generated by the convergence of wavefronts on the surface of a lake. Interference waves find utility in various engineering applications, including but not limited to signal interpreting, musical instruments, as well as radio communication (Papp et al., 2021).

Wave interactions are changing phenomena that manifest when waves come into contact with obstacles, limits, or other waves. There are various categories of wave interactions that can be identified, which encompass:

- *Reflection*: The phenomenon of reflection takes place when waves come across a surface or boundary, resulting in their subsequent rebound back into the initial medium. According to the law of reflection, the angle of incidence is equivalent to the angle of reflection. This phenomenon can be observed in the reflection of sound waves from walls, the reflection of light waves from mirrors, and the reflection of water waves from piers.
- *Refraction*: Refraction is the phenomenon in which waves undergo a change in direction as they transition from a certain medium

to another characterized by distinct properties. The relationship among the angles of incidence and refraction is described by Snell's law. The phenomenon of refraction is frequently observed when light waves undergo a change in direction as they traverse media such as glass or water, resulting in notable effects such as the formation of rainbows (Campos et al., 1990).

- *Diffraction*: Diffraction refers to the phenomenon in which waves undergo bending and spreading as they encounter obstacles or pass through narrow apertures. The phenomenon arises when waves come into contact with obstacles that are of a similar magnitude to their wavelength. Diffraction, an optical phenomenon, plays a pivotal role in the formation of interference patterns. It is particularly significant in various phenomena, such as the propagation of sound around obstacles and the characteristics of light waves when they traverse narrow apertures.

The properties of waves are fundamental attributes that determine the behavior and characteristics of these phenomena. The fundamental characteristics of waves encompass (Couder & Fort, 2006):

- *Amplitude*: The amplitude of a wave refers to the utmost displacement or distance that the wave undergoes from its state of equilibrium. Within the realm of sound waves, the amplitude serves as the determinant of sound loudness, whereas in the domain of light waves, it exerts an influence on the luminous intensity of the light.
- *Wavelength*: The wavelength of a wave can be defined as the spatial separation among two consecutive points that exhibit phase coherence, such as neighboring crests or troughs. The relationship between wavelength and frequency is inversely proportional, indicating that waves with a shorter wavelength exhibit higher frequencies, while waves with longer wavelengths correspond to lower frequencies (Lyu et al., 2018).
- *Frequency*: The frequency of a wave refers to the quantity of oscillations or cycles it undergoes within a given time interval. The unit of measurement for frequency is hertz (Hz), where one hertz corresponds to one cycle per second. Higher frequencies are associated with waves that exhibit more rapid oscillations.
- *Speed*: The velocity of a wave is the measure of the speed at which the wavefront propagates within a given medium. The

calculation of wave speed is contingent upon the characteristics of the medium in which the wave travels, and can be obtained by multiplying the wavelength by the frequency (Choi et al., 2006).

- In summary, the phenomena of wave behavior as well as interference are fundamental principles that hold significant importance in a wide range of both organic and engineered systems. Interference waves arise as a consequence of the phenomenon known as wave superposition, wherein waves combine and interact with each other, resulting in either constructive or destructive interference. Various forms of wave interactions, namely reflection, refraction, as well as diffraction, manifest themselves when waves come into contact with boundaries or obstacles. The comprehension of wave characteristics, encompassing amplitude, wavelength, frequency, as well as speed, enables us to articulate and examine wave phenomena across various domains, such as acoustics, optics, communication, and engineering (Chen et al., 2016).

5.2. WAVE INTERACTIONS

Waves are dynamic entities that engage in interactions with their surrounding environment, resulting in intriguing phenomena like reflection, refraction, as well as diffraction. This section will comprehensively examine wave interactions, encompassing the governing laws and practical implications across diverse facets of our everyday existence (Zhadobov et al., 2011).

5.2.1. Reflection of Waves

The reflection of waves is an essential phenomenon characterized by the interaction of waves with a surface or boundary, resulting in their subsequent rebound back into the medium from which they originated (Figure 5.1). When waves experience reflection, they alter their trajectory without undergoing absorption or transmission across the boundary. The angle of incidence, denoted as the angle formed between the incident wave as well as the normal to the reflecting surface, is equivalent to the angle of reflection, which represents the angle formed between the reflected wave and the normal.

Reflection is a fundamental phenomenon observed in a wide range of organic and engineered systems, encompassing phenomena such as the generation of echoes in sound waves and the development of images in reflections for light waves. The concept of wave interaction with the environment is crucial in the field of wave behavior research, as it enables comprehension and prediction of such phenomena (Bérenger, 1999).

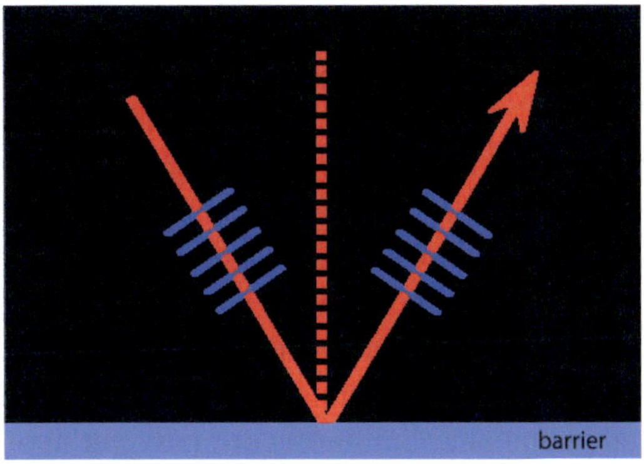

Figure 5.1. Schematic of the reflection of waves (Source: CK-12, Creative Commons License).

5.2.1.1. Law of Reflection

Reflection is a phenomenon that occurs when waves interact with an edge or interface and subsequently rebound back into the medium from which they originated. According to the principle of reflection, the angle of incidence, which refers to the angle formed among the incident wave as well as the perpendicular line to the surface, is equivalent to the angle of reflection, which denotes the angle formed among the reflected wave as well as the perpendicular line. This legislation is relevant to a range of wave phenomena, encompassing light waves, sound waves, as well as water waves (Millikan, 1923).

The phenomenon of wave reflection occurs due to the interaction between waves and an interface that delineates two distinct media. When waves encounter a boundary, a portion of their energy is transferred to the second medium, while the rest of the energy returns back into the initial medium. The measurement of the angle of incidence as well as the angle of reflection is conducted with respect to the normal, a hypothetical line that is perpendicular to the interface (Liu et al., 2016).

5.2.1.2. Examples of Wave Reflection

- ***Sound Waves***: Reflection of sound waves occurs in a multitude of commonplace situations. When an individual vocalizes loudly in the presence of a rigid barrier, the acoustic waves undergo reflection, resulting in the perception of an auditory phenomenon known as an echo. Likewise, within expansive environments like auditoriums and concert halls, sound waves possess the ability to undergo reflection upon encountering walls, ceilings, as well as floors, thereby exerting an impact on the overall auditory encounter (Kaji & Okazaki, 1970).

- ***Light Waves*:** The phenomenon of light wave reflection serves as the fundamental mechanism underlying our visual perception of objects. When incident light encounters a mirror as well as a smooth surface, it undergoes reflection in a precise and predictable manner, facilitating our perception of the surrounding environment. The principle of reflection finds application in the field of optics for the purpose of designing a range of devices, including telescopes and periscopes.
- ***Water Waves*:** The phenomenon of wave reflection is observed when water waves come into contact with a solid obstruction, like a pier or a seawall. The incident waves undergo reflection upon encountering the structure, resulting in the formation of interference patterns and the emergence of an intricate wave pattern (Chang et al., 2001).

5.2.2. Refraction of Waves

The phenomenon of wave refraction is a captivating occurrence that transpires when waves traverse between two distinct mediums characterized by varying properties, such as differences in density as well as refractive index (Figure 5.2). When waves undergo a transition from one medium to another, they exhibit alterations in both their direction and velocity, resulting in a phenomenon known as bending. The degree of bending is contingent upon the angle of incidence as well as the characteristics of the two mediums in question.

Snell's law elucidates the correlation among the angles of refraction and incidence within a specific arrangement of mediums. Refraction is accountable for a multitude of natural occurrences, including the deflection of light when it transitions into water, resulting in the perceptual effect of objects appearing nearer or displaced. Moreover, it assumes a pivotal function in optical instruments such as lenses and prisms, wherein meticulous management of refraction facilitates the creation and manipulation of images (Bordiga et al., 2019).

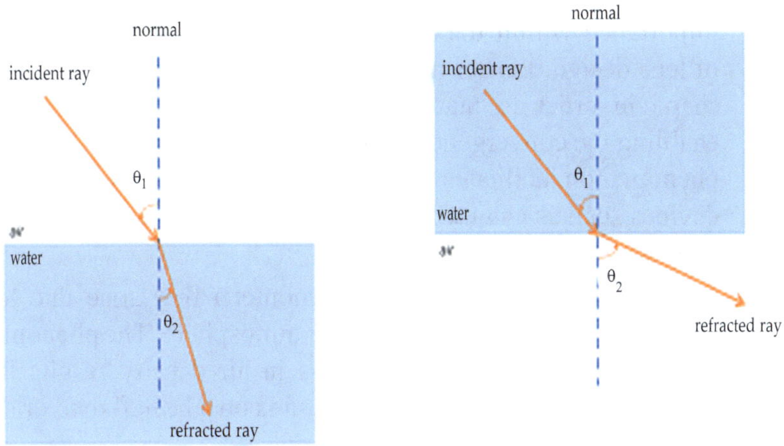

Figure 5.2. Representation of wave refraction (Source: Rochelle Enrera, Creative Commons License).

5.2.2.1. Snell's Law

The phenomenon of refraction is observed when waves transition from a single medium to another, resulting in a change in their velocity as a consequence of disparities in the characteristics of the two media. The alteration in velocity induces a phenomenon wherein the waves undergo deflection or alteration of their trajectory at the interface. The phenomenon in question is described by Snell's law, which is also referred to as the law of refraction (Mulkers et al., 2018).

Snell's law states that the relationship between the sine of the angle of incidence and the sine of the angle of refraction remains constant when considering a specific pair of media. The mathematical representation of Snell's law is as follows:

$$n_1 \sin \theta_1 = n_2 \sin \theta_2$$

where: n_1 and n_2 are the refractive indices of the first and second media, respectively; θ_1 is the angle of incidence, and θ_2 is the angle of refraction (Tanabe et al., 2014).

5.2.2.2. Applications of Wave Refraction

- ***Optics:*** The phenomenon of refraction holds significant importance within the field of optics, particularly in the realm of lens design. Lenses are specifically designed with a particular shape in order to manipulate the refraction of light, thereby enabling the convergence or divergence of light rays. This optical phenomenon facilitates the formation of images in various optical devices such as cameras, telescopes, as well as eyeglasses (Ito & Tanimoto, 1972).
- ***Mirages:*** Mirages are visual phenomena that arise due to the refraction of light within the Earth's atmosphere. The phenomenon of light bending due to variations in air density results in the visual perception of water-like features on a heated road or desert terrain.
- ***Prism Dispersion:*** The phenomenon of refraction in prisms is responsible for the dispersion of light into its individual constituent colors, resulting in visually captivating spectrums reminiscent of rainbows. The occurrence being referred to is commonly known as dispersion and is extensively employed in the field of spectroscopy for the purpose of analyzing light sources (Zetterstrom et al., 2019).

5.2.3. Diffraction of Waves

The phenomenon of wave diffraction is a captivating occurrence in which waves undergo bending and dispersion as they interact with obstacles or traverse narrow apertures. The observed behavior can be attributed to the phenomenon of wavefronts colliding with the boundaries of the barriers or aperture, leading to a subsequent alteration in the direction of the wave. The degree of diffraction is contingent upon the relative proportions between the size of the barriers or aperture and the wavelength of the wave. When the dimensions of the obstacle or slit are of the same order of magnitude as the wavelength, notable diffraction phenomena take place, resulting in the emergence of interference patterns. Diffraction phenomena can be observed in a wide range of natural and engineered systems, encompassing instances such as the propagation of sound within corners, the diffraction of light via narrow slits, and the characteristic behavior of waves when encountering obstacles in their trajectory (Figure 5.3). A comprehensive comprehension of diffraction is imperative for the accurate interpretation of wave phenomena and holds significant importance in various disciplines such as acoustics, optics, as well as radio communication (Levin, 1978).

Tip

Diffraction is a wave behavior that occurs when waves encounter an obstacle or aperture, causing them to bend and spread around the edges. This phenomenon is essential in understanding how waves interact with various structures and openings.

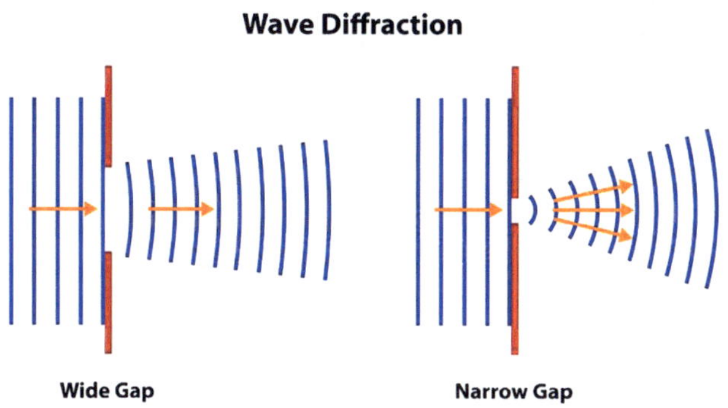

Figure 5.3. Diffraction waves through gap sizes (Source: Dreamstime, Creative Commons License).

5.2.3.1. Diffraction Patterns

Diffraction refers to the phenomenon wherein waves undergo bending and spreading when they come into contact with an obstacle or a slit that is of similar dimensions to the wavelength of the waves. In contrast to the phenomena of reflection as well as refraction, which are characterized by interactions occurring at surfaces, diffraction arises as a consequence of wave interference.

The phenomenon of wave diffraction occurs when waves propagate through a constricted aperture or encounter an obstacle, resulting in their bending and subsequent dispersion, thereby giving rise to a distinctive diffraction pattern. An instance of this phenomenon occurs when light traverses a narrow aperture, resulting in the emergence of a sequence of illuminated and shadowed regions on a surface situated beyond the aperture (Tonner et al., 1991).

5.2.3.2. Importance of Diffraction in Everyday Life

- ***Hearing***: Diffraction in the context of sound waves enables the perception of auditory stimuli emanating from regions obstructed by corners or obstacles. The phenomenon of sound waves exhibiting the ability to undergo bending around physical objects, thereby enabling their reception by our auditory system despite the presence of obstructions that impede direct line of sight, has been observed (Miller, 1974).
- ***Radio Waves***: Diffraction plays a crucial role in the realm of radio communication, particularly in the context of broadcasting radio signals. In urban environments, radio waves have the ability to undergo diffraction, allowing them to propagate around buildings as well as natural obstacles. This phenomenon contributes to improved coverage as well as reception in such areas (Hewish, 1952).
- ***Microscopy***: Diffraction constitutes a fundamental aspect within the field of microscopy. The resolution of optical microscopes is constrained due to the phenomenon of diffraction, wherein light waves exhibit bending as they encounter the edges of minute objects, thereby causing image blurring.

The phenomenon of wave interactions holds considerable importance in influencing our understanding of the surrounding environment. The phenomena of reflection, refraction, as well as diffraction are intriguing occurrences that arise when waves interact with boundaries or obstacles. The phenomenon of wave reflection is governed by the law of reflection, which outlines the principles dictating the behavior of waves as they rebound off surfaces.

On the other hand, the bend of waves during the process of refraction is elucidated by Snell's law. In contrast, diffraction arises from the phenomenon of wave interference as well as gives rise to the phenomenon of wave bending and dispersion when waves traverse narrow apertures or encounter obstructions. Comprehending these wave interactions holds paramount importance for a multitude of applications in the fields of optics, communication, and daily existence, thereby enhancing our comprehension of wave behavior and the underlying physics of our surroundings (Miao et al., 2004).

5.3. WAVE INTERFERENCE

The phenomenon of wave interference is a captivating occurrence that arises from the superposition or overlapping of two or more waves (Figure 5.4). When waves intersect, their respective displacements combine, leading to a resultant displacement at every overlapping point. This phenomenon results in either the reinforcement or cancellation of the waves, contingent upon their relative phases as well as amplitudes.

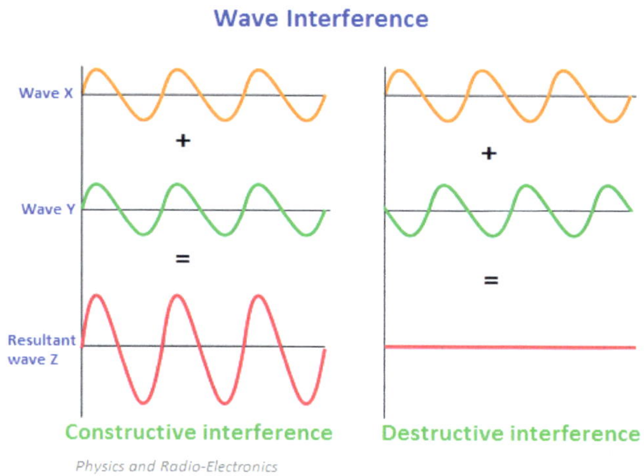

Figure 5.4. Schematic of the wave interference (Source: Physics and Radio-Electronics, Creative Commons License)

Interference phenomena can manifest in various wave types, encompassing sound waves, waves of light, water waves, as well as electromagnetic waves. The underlying principle of interference is derived from the concept of superposition, which posits that the combined displacement resulting from the meeting of two or more waves at a specific point is equal to the algebraic sum of their respective individual shifts at that point (Eibenberger et al., 2013).

The principle of superposition enables the mathematical characterization of wave behavior in the context of interference. When two waves, as described by their respective wave equations, intersect at a specific location, the cumulative shifting of the wave that results at said location is equal to the summation of the movements of each of the waves.

The total displacement, denoted as y, at any given point and time t can be expressed as a function of the amplitudes A_1 as well as A_2, wavelengths λ_1 as well as λ_2, and angular frequencies ω_1 as well as ω_2.

$$y = A_1 \sin(\omega_1 t - k_1 x) + A_2 \sin(\omega_2 t - k_2 x)$$

In the given context, let x denote the spatial coordinate along the direction of the wave. Additionally, k_1 and k_2 represent the wave numbers, which are directly associated with the wavelengths through the equation $k = 2\pi/\lambda$ (Yeung & Sphaier, 1989).

5.3.1. Types of Interference

The categorization of interference pertains to the various modalities through which waves engage with one another upon encountering or superimposing. Interference can be classified into two main categories: constructive interference and destructive interference. Constructive interference occurs when waves of identical frequency as well as phase align in such a way as to strengthen one another, leading to an amplified combined amplitude. This phenomenon results in the formation of areas characterized by increased intensity and brightness in the context of light waves, and amplified volume in the context of sound waves. Conversely, within the context of destructive interference, waves exhibiting opposite phases undergo mutual annihilation, resulting in a decrease in amplitude. This phenomenon results in areas of reduced intensity or diminished brightness in the context of light waves, and decreased volume or attenuated amplitude in the context of sound waves. A comprehensive comprehension of these forms of interference holds significant importance in multiple disciplines, such as optics, acoustics, as well as radio communications, as they govern the manner in which waves interact and impact the overall dynamics and characteristics of wave systems (Zhang et al., 2015).

Remember

There are two main types of interference: constructive interference, where waves combine to produce a larger amplitude, and destructive interference, where waves cancel each other out and result in a smaller or zero amplitude. These interference patterns are crucial in many aspects of wave behavior and applications.

5.3.1.1. Constructive Interference

Constructive interference is observed when two waves intersect in phase, resulting in the alignment of their peaks and troughs. When such an occurrence takes place, the amplitudes of the waves combine, leading to the formation of a resultant wave with an increased amplitude. The process of reinforcement results in a heightened level of energy at the locations where two waves overlap, thereby generating areas characterized by constructive interference.

Constructive interference occurs when two or more waves combine to produce a resultant wave with an increased intensity or magnitude compared to the individual waves. This phenomenon is frequently observed in musical instruments, wherein sound waves of identical frequency and in phase amalgamate to generate amplified and more resonant tones.

5.3.1.2. Destructive Interference

Destructive interference is observed when two waves intersect in a state of phase opposition, wherein the crests of one wave coincide with the troughs of the opposing wave. In this particular scenario, the amplitudes undergo subtraction, leading to the emergence of a wave characterized by a diminished amplitude or potentially complete annulment.

During the occurrence of destructive interference, the resultant wave exhibits a diminished intensity or magnitude in comparison to the constituent waves. The aforementioned phenomenon can be noticed with noise-canceling headphones, wherein the detection of external sound waves prompts the generation of an anti-phase sound wave. This anti-phase sound wave serves to counteract and nullify the undesired noise, resulting in a decrease in sound intensity (Polifke et al., 2001).

5.3.2. Coherent and Incoherent Sources

Coherent Sources: In the context of interference, coherence pertains to the correlation between the phases of waves originating from distinct sources. Coherent sources are characterized by their ability to maintain a consistent phase relationship, whereby the phase difference among the waves stays constant throughout the course of time.

When waves originating from coherent sources combine, the resulting interference pattern exhibits a clearly defined and stable nature, resulting in discernible areas of constructive as well as destructive interference. Instances of coherent sources encompass phenomena such as the propagation of waves originating from a solitary light source, the emission of a laser beam, or the operation of two speakers that are stimulated by an identical signal.

Incoherent Sources: Incoherent sources refer to those that lack a consistent phase relationship, leading to variations in the phase distinction between the waves. When waves originating from incoherent sources undergo superposition, the resulting interference pattern exhibits reduced clarity and may exhibit rapid temporal variations (Cadzow, 1988).

5.3.3. Applications of Wave Interference

Following are the applications of wave interference:

5.3.3.1. Interference in Music and Sound

Musicians as well as sound engineers utilize interference in diverse manners to generate captivating auditory experiences. The utilization of beats, which arise from the interference between two frequencies that differ slightly, is a prevalent method employed in the process of tuning musical instruments. When two musical tones with slightly varying frequencies are sounded simultaneously, they generate a beat frequency equivalent to the disparity between their respective frequencies. Musicians possess the ability to fine-tune their instruments by attuning themselves to the auditory cues provided by the beats (Li et al., 2017).

5.3.3.2. Interference in Light Waves

Interference in light waves has significant implications in many optical devices and phenomena.

- ***Thin-Film Interference*:** Thin-film interference is a phenomenon that arises when light waves undergo reflection and transmission upon encountering a thin layer of material. This phenomenon gives rise to the formation of vivid hues in bubbles of soap, oil slicks, as well as anti-reflective coatings found on eyeglasses.
- ***Michelson Interferometer*:** The Michelson interferometer is an optical instrument of high precision that utilizes the phenomenon of interference in order to quantify minute alterations in the length of a given path. The utilization of this technology is prevalent across diverse scientific and engineering domains, encompassing the measurement of the velocity of light and the detection of minute displacements in architectural frameworks.
- ***Holography*:** The technique of holography is predicated upon the utilization of interference patterns generated by laser light in order

to produce three-dimensional visual representations. Holography is a technique that enables the capture of phase information from light waves, facilitating the reconstruction of three-dimensional scenes upon illumination (Shapira et al., 2015).

5.3.3.3. Interference in Radio Waves

The phenomenon of disruption in radio waves presents both benefits and obstacles in the realm of communication technologies.

- ***Antenna Arrays*:** In the realm of radio communication, the utilization of multiple antennas can be employed to generate constructive interference in a designated direction, thereby augmenting the potency of the transmitted signal and ameliorating both the range and quality of communication.
- ***Signal Degradation*:** Signal degradation as well as noise in radio communications can occur due to interference caused by various electronic devices as well as environmental factors. The effective management and implementation of appropriate mitigation strategies are necessary to ensure the reliability of communication in relation to this particular phenomenon.

In summary, wave interference is an intriguing phenomenon that arises from the superposition of two or more waves, resulting in either reinforcement or dismissal of the wave amplitudes. Constructive interference amplifies the amplitude of the resultant wave, whereas destructive interference diminishes it. The establishment of coherence among wave sources is of utmost importance in order to achieve clearly defined interference patterns. The phenomenon of interference finds practical utility in a range of domains such as music, optics, as well as radio communication, thereby enhancing our comprehension of wave characteristics and facilitating diverse technological progressions (Hanada et al., 2001).

5.4. STANDING WAVES

The occurrence of standing waves is a captivating phenomenon wherein two waves of identical frequency and amplitude, propagating in opposing directions, combine to generate a wave pattern that seems motionless or "stationary" within a given space. In contrast to traveling waves which propagate via a medium, stationary waves exhibit oscillatory behavior at a fixed location, retaining their form and magnitude throughout the duration.

The phenomenon of standing waves is determined by the principle of superposition that asserts that the displacements of two waves combine at each point of intersection. When the peak of one wave coincides with the trough of another wave, destructive interference occurs, leading to the formation of nodes, which are locations characterized by minimal displacement. In contrast, when the peaks of the two waves or the valleys of both waves coincide, they exhibit constructive interference, resulting in areas of maximum displacement known as antinodes (Suhayda, 1974).

5.4.1. Nodes and Antinodes

Nodes as well as antinodes are fundamental components of standing waves that have a substantial impact on the formation of the wave pattern (Figure 5.5).

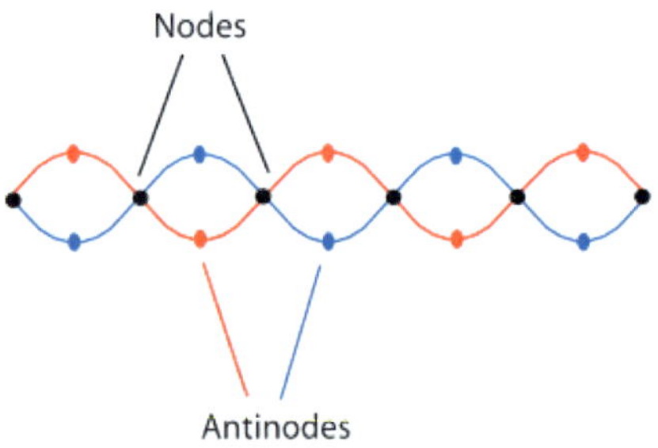

Figure 5.5. A standing wave pattern has nodes and antinodes-two extreme positions of the resonance pattern (Source: Robert G. Sacco, Creative Commons License).

- ***Nodes*:** Nodes are defined as locations within a standing wave where the displacement is zero. At these specific locations, the two waves that have been combined are in a state of complete phase opposition, resulting in their complete cancellation. Consequently, the particles comprising the medium do not undergo any overall displacement, resulting in the wave appearing motionless at these specific locations. In a given mode of a standing wave, the quantity of nodes stays constant.
- ***Antinodes*:** Antinodes represent the locations within a standing wave where the displacement reaches its maximum value. At these specific locations, the two waves that have been combined are in perfect alignment and exhibit constructive interference. The phenomenon of constructive interference leads to the attainment

of the maximum amplitude of oscillation at the points known as antinodes. Similar to nodes, the quantity of antinodes remains consistent for a specific mode of a standing wave (Ospina et al., 2022).

5.4.2. Harmonics and Overtones

Standing waves exhibit distinct frequencies where they can manifest, referred to as harmonics and natural frequencies. The observed harmonics are directly associated with the fundamental frequency as well as its subsequent multiples, commonly referred to as overtones.

- ***Fundamental Frequency*:** The fundamental frequency refers to the minimum frequency whereby a standing wave can be established within a particular medium, subject to specific boundary conditions. The oscillatory system under consideration exhibits a single antinode as well as two nodes, thereby representing the most elementary mode of oscillation. When a guitar string is picked it undergoes vibration at its essential frequency, resulting in the emission of the lowest audible pitch.
- ***Overtones*:** Overtones refer to the occurrence of higher frequencies that can give rise to standing waves, in addition to the basic frequency. The higher modes exhibit a plurality of antinodes and nodes, and manifest at whole number multiples of the basic frequency. The initial overtone, alternatively referred to as the second harmonic, exhibits a configuration of two antinodes as well as three nodes. Successive harmonics exhibit a greater number of antinodes as well as nodes, leading to intricate wave formations and elevated frequencies (Saito & Inoue, 2003).

5.4.3. Practical Applications of Standing Waves

5.4.3.1. Musical Instruments

The phenomenon of standing waves plays a crucial role in the generation of musical sounds across a wide range of instruments. Musical instruments, including those with strings, pipes, and membranes, depend on the establishment of standing waves in order to produce distinct pitches.

- ***String Instruments*:** Musical instruments such as guitars, violins, as well as pianos employ the mechanism of vibrating strings to generate auditory frequencies associated with musical tones. If a string is subjected to plucking or bowing, it initiates the formation of standing waves that encompass multiple harmonics, each of which corresponds to a distinct note or pitch. Musicians possess the ability to manipulate the harmonics and generate diverse musical tones by modifying the length, tension, as well as mass of the string.

- ***Wind Instruments*:** Wind instruments, including flutes, clarinets, as well as organ pipes, employ air columns to generate stationary waves. When air is introduced into the tool, it undergoes oscillation within the tube, resulting in the formation of standing waves characterized by specific harmonics. Musicians possess the ability to manipulate the pitch and generate distinct musical tones by modifying the size and form of the tube (Campbell, 1999).

5.4.3.2. Resonance Phenomena

Resonance is an intriguing phenomenon associated with standing waves, which manifests when an outside force aligns with the inherent frequency of a substance or system. When the frequency of the external force coincides with the inherent frequency of the object, it induces resonance, causing a significant amplification of amplitude and efficient transfer of energy, thereby eliciting a pronounced response.

- ***Acoustic Resonance*:** Acoustic resonance is a phenomenon observed in the fields of music and sound engineering, whereby the frequency of a sound wave aligns with the inherent frequency of a resonant cavity or space. The phenomenon of resonance serves to amplify particular frequencies, thereby augmenting the overall auditory experience and engendering a perception of heightened quality and profoundness within musical renditions.
- ***Mechanical Resonance*:** Resonance can have an impact on mechanical systems, including structures like bridges and buildings, as well as machinery, when external forces align with their inherent frequencies. If not adequately considered in the design and engineering process, this phenomenon has the potential to result in structural damage or failure.

In summary, standing waves are captivating phenomena that manifest when two waves combine, resulting in stationary patterns characterized by nodes and antinodes. The presence of standing wave patterns plays a crucial role in the generation of musical sounds across a range of instruments, as the distinct pitches are determined by the harmonics and overtones. Resonance, which is closely associated with the phenomenon of standing waves, assumes a pivotal role in augmenting sound fidelity and concurrently presents intricate obstacles in the domains of engineering as well as structural design. The comprehension of the attributes and uses of standing waves enhances our understanding of wave phenomena and its pragmatic implications in disciplines such as music, engineering, and diverse areas of study (Rudenko, 2009).

5.5. BEATS AND BEAT FREQUENCIES

The phenomenon of beats is a captivating acoustic occurrence that arises from the interference between two sound waves with slightly disparate frequencies. When the phenomenon of wave superposition occurs, it gives rise to periodic fluctuations in the collective amplitude, leading to the sense of a pulsating and beating sound. The understanding of beats holds significant importance in the fields of music and engineering, as it offers important insights into the dynamics of sound waves as well as their practical implications (Smallman et al., 1984).

5.5.1. Formation and Characteristics of Beats

The phenomenon of beats occurs when two sound waves, characterized by slightly disparate frequencies, propagate through a common medium and superpose (Figure 5.6). The frequencies of both waves exhibit a lack of significant disparity, thereby precluding their classification as distinct sounds. However, their discernible dissimilarity is sufficient to generate an audible pattern of interference.

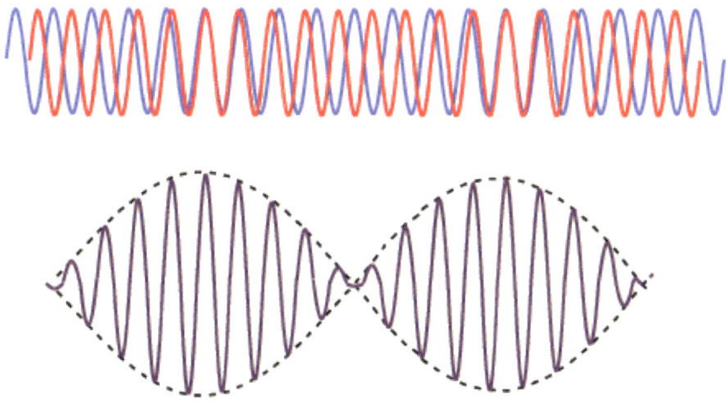

Figure 5.6. Beats are caused by the interference of two waves at the same point in space. This plot of the variation of resultant amplitude with time shows the

periodic increase and decrease for two sine waves (Source: Hyper Physics, Creative Commons License).

The phenomenon of beats can be comprehended by examining the interaction between two sound waves characterized by frequencies f_1 and f_2, with f_1 being greater than f_2. When these waves undergo superposition, their instantaneous amplitudes combine. During specific instances, the two waves exhibit phase alignment, leading to constructive interference and increased combined amplitudes. During different temporal instances, the waves exhibit a phase difference, resulting in the occurrence of destructive interference and subsequently reduced combined amplitudes.

The temporal fluctuations in the aggregate magnitude give rise to the auditory sensation of a pulsating sound, characterized by a recurrent sequence of high and low intensity segments. The beat frequency, represented by the symbol f_beat, is calculated as the numerical disparity among the frequencies of both waves:

$$f\_beat = |f_1 - f_2|$$

The beat frequency is a measure of the frequency where the amplitude of the resultant wave oscillates among its highest and lowest values. The occurrence of beats will be faster as the frequency difference between the two waves increases (Vorob'ev et al., 1981).

5.5.2. Mathematical Expression of Beat Frequency

The mathematical formulation of the beat frequency serves to measure the phenomenon and facilitate comprehension of its inherent characteristics. When two waves of sound with frequencies f_1 as well as f_2 superpose, the resultant wave can be mathematically expressed as:

$$y(t) = A_1 \sin(2\pi f_1 t) + A_2 \sin(2\pi f_2 t)$$

The variable y(t) represents the instantaneous shifting of the wave at a given time t, while A_1 and A_2 denote the amplitudes of the respective individual waves.

Utilizing the trigonometric identity $\sin(A) + \sin(B) = 2 \sin((A + B)/2) \cos((A - B)/2)$, it is possible to represent the function y(t) as:

$$y(t) = 2A \sin(2\pi((f_1 + f_2)/2)t) \cos(2\pi((f_1 - f_2)/2)t)$$

The beat frequency, denoted as f_beat, can be expressed as $2\pi((f_1 - f_2)/2)t$, that can be further simplified to $\pi(f_1 - f_2)t$. Therefore, the beat frequency can be expressed as:

$f\_beat = |f_1 - f_2|$

The perception of beats in sound is significantly influenced by the amplitude modulation resulting from the beat frequency. The frequency of the amplitude oscillation governs the tempo whereby the beats are perceived, while the disparity in amplitude between the highest and lowest values impacts the intensity of the beat's auditory perception (Bondarenko, 2010).

5.5.3. Applications of Beats in Music and Engineering

5.5.3.1. Music

Beats possess practical applications within the realm of music and are employed in diverse manners by musicians as well as composers.

- ***Tuning Instruments***: The utilization of beats is a prevalent practice in the process of tuning musical instruments. When two musical tones are played concurrently, with frequencies that are in close proximity but not precisely identical, the phenomenon of beats can be perceived. Musicians employ techniques such as modifying the tension and length of strings, adjusting the air column in wind-based instruments, or manipulating the frequency of digital instruments in order to eliminate beats, thereby indicating the fact that both notes are harmoniously tuned (Kornysheva et al., 2010).
- ***Harmonics and Overtones***: Beats arise as a result of the interference that occurs when various harmonics as well as overtones of musical devices interact. The distinct auditory characteristics and timbral qualities of different instruments are attributed to the distinctive amalgamation of harmonics they produce.
- ***Creating Musical Effects***: Musicians possess the intentional capacity to incorporate beats into compositions with the purpose of generating distinctive musical effects. This technique is prevalent in specific genres, including minimalism, surrounding music, and certain avant-garde styles (Griffin et al., 2010).

5.5.3.2. Engineering

Beats possess practical applications in the realm of engineering and other diverse fields that extend beyond the domain of music.

- ***Audio Signal Processing*:** Within the field of audio engineering, beats serve as a fundamental component in signal processing methodologies aimed at generating various audio effects, including but not limited to tremolo and vibrato. The aforementioned effects serve to modulate either the amplitude or frequency of the sound wave, thereby generating a sound that is characterized by its expressive and dynamic qualities.
- ***Frequency Measurement*:** The utilization of beats is observed in frequency measurement instruments, like beat frequency oscillators. Engineers can accurately ascertain the unknown frequency by conducting a comparative analysis of the beat frequency generated by two signals, one possessing a known frequency and the other an unknown frequency.
- ***Mechanical Vibrations*:** Beats may be noticed in vibrating systems within the field of mechanical engineering, including machinery, engines, as well as structures. In order to maintain the structural integrity of various constructions, engineers must take into consideration the phenomena of resonance as well as potential beat frequencies, as these can lead to detrimental vibrations.

The phenomenon of beats is a captivating acoustic occurrence that arises from the interference between waves of sound of slightly disparate frequencies. The sensation of a pulsating and beating sound arises from the periodic fluctuation in the combined amplitudes of the waves. The quantification of the phenomenon is facilitated by the mathematical representation of the beat frequency, which in turn offers valuable insights as well as practical applications in the fields of music and engineering. Musicians employ beats in order to calibrate instruments and generate distinctive musical embellishments, whereas engineers harness them for the purpose of quantifying frequencies and manipulating signals. Gaining comprehension enhances our understanding of wave phenomena (Langley et al., 2009).

5.6. INTERFERENCE IN MODERN TECHNOLOGY

Interference is a captivating natural occurrence that holds significant importance in contemporary technology. The principles of interference have been effectively utilized by engineers and scientists to create cutting-edge technologies that have had a transformative impact on communication, measurement, and numerous other domains. This section will examine two notable applications of interference in contemporary technology: fiber-optic communication as well as interferometry in metrology (Siddiqui et al., 2021).

5.6.1. Fiber-Optic Communication

Fiber-optic communication is a technological system that employs optical fibers as a medium for transmitting data in the form of light pulses. The technology in question has emerged as a fundamental component of contemporary communication systems, facilitating the efficient transmission of data at high speeds across extensive distances while minimizing the degradation of signals.

Did You Know?

Fiber-optic communication is a technology that utilizes the principle of total internal reflection in optical fibers to transmit data as pulses of light. This allows for high-speed and long-distance transmission of information, making it a critical component of modern telecommunication systems.

Interference is a crucial factor in the field of fiber-optic communication, as it plays a pivotal role in the implementation of multiplexing. Multiplexing is a technique employed to enable the simultaneous transmission of multiple signals over a single fiber. There are two primary techniques of multiplexing that are based on the principle of interference (Lam et al., 2010):

- ***Wavelength Division Multiplexing (WDM)*:** Wavelength-division multiplexing (WDM) is a technique that leverages the phenomenon of interference among distinct wavelengths of light in order to concurrently transmit several signals over a single optical fiber. A distinct wavelength is allocated to each signal, and these signals are consolidated into a unified optical signal through the utilization of a multiplexer. At the point of reception, a demultiplexer is utilized to segregate the signals according to their respective wavelengths.
- ***Coherent Optical Communication:*** The process of coherent optical communication involves the utilization of interference phenomena to effectively detect and decode both the amplitude and phase information contained within light signals. The system utilizes coherent receivers capable of extracting information from

interference patterns, thereby enabling increased data rates and extended transmission distances.

Advantages:

- *High Capacity*: Fiber-optic communication provides a wide bandwidth, enabling the concurrent transmission of substantial volumes of data.
- *Low Signal Loss*: Optical fibers exhibit negligible signal attenuation, thereby enabling the transmission of data over extensive distances with minimal degradation.
- *Security*: Fiber-optic communication exhibits a higher degree of resistance against interception and eavesdropping in contrast to conventional copper-based communication systems (Joshi et al., 2005).

5.6.2. Interferometry in Metrology

Interferometry is a highly accurate measurement methodology that utilizes the principle of interference in order to ascertain minute variations in length, displacement, and refractive index. The utilization of this technique is prevalent in the field of metrology as well as scientific research as a means to attain measurements of exceptional precision.

The Utilization of Interference in Various Contexts

Interferometry is a technique that entails the division of a light and electromagnetic wavefront into multiple paths, enabling them to traverse distinct distances prior to being reunited. The resultant interference pattern yields insights into the relative phase and path disparity among the waves, which is directly linked to the measured physical quantity.

Several prevalent forms of interferometry include:

- ***Michelson Interferometer:*** The Michelson interferometer is widely recognized as one of the most prominent categories of interferometers. A beam splitter is employed to bifurcate the incident light into two distinct trajectories, wherein one trajectory undergoes reflection upon impinging on a mirror that is capable of being displaced. Variations in the path length lead to a displacement of the interference pattern, enabling the derivation of accurate measurements for parameters such as length and refractive index.

- ***Fizeau Interferometer*:** The Fizeau interferometer employs a partially reflective plate or lens in order to generate interference fringes. Surface profiling as well as testing of optical components frequently employ this technique.

Advantages:

- *High Precision*: Interferometry is a highly precise measurement technique that holds immense value in scientific research as well as industries that prioritize accuracy.
- *Non-Contact*: Interferometric measurements are characterized by their non-contact nature, wherein the measurement system does not establish physical contact with the object under examination. This particular attribute proves advantageous when handling materials that are delicate or sensitive (Mokdad et al., 2004).

5.6.3. Challenges and Solutions in Managing Interference

Interference, despite its manifold advantageous applications, poses inherent challenges that necessitate resolution within diverse technological systems.

5.6.3.1. Interference Mitigation Techniques

In scenarios where the presence of interference can lead to a decline in signal quality or compromise the integrity of data, diverse strategies are implemented to alleviate its impact.

- ***Shielding*:** Shielding is a widely employed method utilized to mitigate the impact of external electromagnetic interference on vulnerable electronic devices. The internal components of a device or system can be safeguarded by enveloping them within a conductive material, which effectively redirects electromagnetic fields away from the enclosed area.
- ***Filtering*:** Filters are employed for the purpose of selectively reducing the amplitude of particular frequency ranges, thereby permitting only the wanted signals to propagate while impeding undesired interference.
- ***Frequency Planning*:** Effective frequency planning is crucial in systems for wireless communication as it serves to mitigate interference among various channels and optimize the utilization of the accessible spectrum (Bouras et al., 2014).

5.6.3.2. Avoiding Signal Degradation

The presence of interference within communication systems can result in a decline in signal quality, thereby diminishing the reliability and overall effectiveness of data transmission. In order to mitigate the effects of signal degradation, a range of strategies are employed (Kaleem et al., 2014):

- ***Redundancy and Error Correction*****:** Within the realm of communication systems, the implementation of redundancy as well as error correction techniques serves the purpose of identifying and rectifying errors that arise as a result of interference. Through the process of encoding supplementary information within the transmitted data, the recipient is able to discern and rectify errors, thereby safeguarding the integrity of the data (Prévost et al., 2012).
- ***Signal Processing Algorithms*****:** Sophisticated signal processing algorithms are utilized to effectively eliminate undesired noise and interference, thereby enhancing the overall quality of received signals.
- ***Beamforming*****:** Beamforming is a technique employed in wireless communication to concentrate the sending and receiving of signals in a targeted direction, thereby mitigating interference originating from alternative directions.

Interference plays a pivotal role in contemporary technology, facilitating the implementation of sophisticated applications in the domains of communication and metrology. Fiber-optic communication utilizes interference for the purpose of multiplexing as well as coherent detection, thereby enabling the efficient transmission of data over extended distances with high speed and reliability. Interferometry plays a crucial role in metrology as well as scientific research, offering a means to achieve highly accurate measurements. Consequently, it has become indispensable across a range of industries. Interference presents significant obstacles in the management of signal degradation and information corruption. However, scientists and engineers employ a diverse array of techniques to effectively mitigate these challenges, thereby guaranteeing the dependability and effectiveness of contemporary technological systems. The comprehension and utilization of interference persistently propel advancements and enhancements in diverse domains, enhancing our capacities and expanding the frontiers of contemporary technology (Kulkarni et al., 2020).

Activity 5.1

In an acoustic concert hall, sound diffraction and reflection play significant roles in creating a high-quality audio experience for the audience. Describe how diffraction and reflection of sound waves can be strategically managed in the design of the concert hall to enhance the overall acoustics and ensure an enjoyable listening experience for the audience.

SUMMARY

The chapter titled "Wave Behavior and Interference" delves into the intricate realm of waves and their interactions within various natural and engineered systems. The text commences by providing an introduction to the notion of waves, elucidating their inherent properties, and emphasizing the significance of comprehending wave dynamics across diverse disciplines. The chapter explores the phenomenon of interference waves, wherein the superposition of two or more waves results in either constructive or destructive interference. Constructive interference leads to the formation of areas characterized by increased amplitude, whereas destructive interference gives rise to regions characterized by decreased amplitude. Subsequently, the chapter delves into an examination of diverse forms of wave interactions, encompassing reflection, refraction, as well as diffraction. Reflection is the phenomenon in which waves undergo a change in direction upon encountering surfaces or boundaries, resulting in their bouncing back. On the other hand, refraction refers to the bending of waves as they transition from one medium to another. In contrast, diffraction refers to the phenomenon in which waves undergo bending and spreading as they encounter obstacles or pass through narrow apertures.In addition, the chapter emphasizes the fundamental properties of waves, namely amplitude, wavelength, frequency, as well as speed. The aforementioned properties delineate the characteristics and fundamental attributes of waves, exerting a substantial influence on their utilization in various domains such as acoustics, optics, telecommunications, and engineering. The chapter additionally examines the practical implementations of interference in contemporary technology, including fiber-optic communication. This technology effectively employs interference for multiplexing as well as coherent detection purposes. Moreover, the utilization of interferometry in the field of metrology enables the attainment of exceptionally accurate measurements. Finally, this chapter discusses the various difficulties encountered in the management of interference, encompassing strategies for interference mitigation and the utilization of algorithms for signal processing to prevent signal deterioration. In summary, the publication titled "Wave Behavior and Interference" offers a thorough examination of the characteristics of waves, their interactions with one another, and the captivating occurrence known as interference. This elucidates the application of these concepts across various disciplines, encompassing communication and metrology, thereby augmenting our comprehension of wave behavior as well as its pragmatic implications in contemporary technology and daily existence.

REVIEW QUESTIONS

1. What are the three main wave interactions discussed in this chapter? Describe the processes of reflection, refraction, and diffraction of waves.
2. Explain the concept of wave interference and its significance in wave behavior. Provide examples of the types of interference.
3. Differentiate between coherent and incoherent sources of waves. How does the coherence of sources affect the interference pattern?
4. Discuss the applications of wave interference in various fields. How is wave interference utilized in practical scenarios?
5. Define standing waves and identify the locations of nodes and antinodes in such waves. How do harmonics and overtones relate to standing wave patterns?
6. What are beats and beat frequencies in the context of wave behavior? How can the beat frequency be mathematically expressed?
7. Explore the applications of beats in music and engineering. How are beats utilized to enhance or manage sound and wave signals?
8. Describe two examples of interference in modern technology, focusing on fiber-optic communication and interferometry in metrology. What challenges arise from interference, and how are they addressed?

REFERENCES

1. Bérenger, J. P. (1999). Evanescent waves in PML's: Origin of the numerical reflection in wave-structure interaction problems. *IEEE Transactions on Antennas and Propagation*, *47*(10), 1497–1503.
2. Bondarenko, E. A. (2010). Expressions for counterpropagating waves beat frequency of uniformly rotating laser gyro. *IEEE Central Ukraine Joint Chapter "Methods and Systems of Navigation and Motion Control"(MSNMC-2010). Kyiv, Ukraine*, 54(52), 13–16.
3. Bordiga, G., Cabras, L., Piccolroaz, A., & Bigoni, D. (2019). Prestress tuning of negative refraction and wave channeling from flexural sources. *Applied Physics Letters*, *114*(4), 987–990.
4. Bouras, C., Diles, G., Kokkinos, V., Kontodimas, K., & Papazois, A. (2014). A simulation framework for evaluating interference mitigation techniques in heterogeneous cellular environments. *Wireless Personal Communications*, *77*, 1213–1237.
5. Cadzow, J. A. (1988). A high resolution direction-of-arrival algorithm for narrow-band coherent and incoherent sources. *IEEE Transactions on Acoustics, Speech, and Signal Processing*, *36*(7), 965–979.
6. Campbell, D. M. (1999). Nonlinear dynamics of musical reed and brass wind instruments. *Contemporary Physics*, *40*(6), 415–431.
7. Campos, R. A., Saleh, B. E., & Teich, M. C. (1990). Fourth-order interference of joint single-photon wave packets in lossless optical systems. *Physical Review A*, *42*(7), 4127.
8. Chang, K. A., Hsu, T. J., & Liu, P. L. F. (2001). Vortex generation and evolution in water waves propagating over a submerged rectangular obstacle: Part I. Solitary waves. *Coastal Engineering*, *44*(1), 13–36.
9. Chen, X. K., Xie, Z. X., Zhou, W. X., Tang, L. M., & Chen, K. Q. (2016). Phonon wave interference in graphene and boron nitride superlattice. *Applied Physics Letters*, *109*(2), 752–755.
10. Choi, S., Lee, K. S., & Kim, S. K. (2006). Spin-wave interference. *Applied Physics Letters*, *89*(6), 654–656.
11. Couder, Y., & Fort, E. (2006). Single-particle diffraction and interference at a macroscopic scale. *Physical Review Letters*, *97*(15), 154101.
12. Donati, O., Missiroli, G. P., & Pozzi, G. (1973). An experiment on electron interference. *American Journal of Physics*, *41*(5), 639–644.

13. Eibenberger, S., Gerlich, S., Arndt, M., Mayor, M., & Tüxen, J. (2013). Matter–wave interference of particles selected from a molecular library with masses exceeding 10000 amu. *Physical Chemistry Chemical Physics*, *15*(35), 14696–14700.
14. Griffin, G., Kim, Y. E., & Turnbull, D. (2010, March). Beat-sync-mash-coder: A web application for real-time creation of beat-synchronous music mashups. In *2010 IEEE International Conference on Acoustics, Speech and Signal Processing.* IEEE. (Vol. 2, pp. 437–440).
15. Hanada, E., Kodama, K., Takano, K., Watanabe, Y., & Nose, Y. (2001). Possible electromagnetic interference with electronic medical equipment by radio waves coming from outside the hospital. *Journal of Medical Systems*, *25*, 257–267.
16. Hewish, A. (1952). The diffraction of galactic radio waves as a method of investigating the irregular structure of the ionosphere. *Proceedings of the Royal Society of London. Series A. Mathematical and Physical Sciences*, *214*(1119), 494–514.
17. Ito, Y., & Tanimoto, K. (1972). Method of Numerical Analysis of Wave Propagation—Application to Wave Diffraction and Refraction—. In *Coastal Engineering 1972* (pp. 503–522).
18. Joshi, A., Wang, X., Mohr, D., Becker, D., & Wree, C. (2005, May). Balanced photoreceivers for analog and digital fiber optic communications. In *Enabling Photonics Technologies for Defense, Security, and Aerospace Applications.* SPIE. (Vol. 5814, pp. 39–50).
19. Kaji, S., & Okazaki, T. (1970). Propagation of sound waves through a blade row: I. Analysis based on the semi-actuator disk theory. *Journal of Sound and Vibration*, *11*(3), 339–353.
20. Kaleem, Z., Hui, B., & Chang, K. (2014). QoS priority-based dynamic frequency band allocation algorithm for load balancing and interference avoidance in 3GPP LTE HetNet. *EURASIP Journal on Wireless Communications and Networking*, *2014*(1), 1–18.
21. Kornysheva, K., von Cramon, D. Y., Jacobsen, T., & Schubotz, R. I. (2010). Tuning-in to the beat: Aesthetic appreciation of musical rhythms correlates with a premotor activity boost. *Human Brain Mapping*, *31*(1), 48–64.
22. Kulkarni, V., Narayana, K., & Sahoo, S. K. (2020). A Survey on Interference Avoiding Methods for Wireless Sensor Networks Working in the 2.4 GHz Frequency Band. *Journal of Engineering Science &*

Technology Review, *13*(3), 654–656.

23. Lam, C. F., Liu, H., Koley, B., Zhao, X., Kamalov, V., & Gill, V. (2010). Fiber optic communication technologies: What's needed for datacenter network operations. *IEEE Communications Magazine*, *48*(7), 32–39.
24. Langley, P., Bowers, E. J., & Murray, A. (2009). Principal component analysis as a tool for analyzing beat-to-beat changes in ECG features: application to ECG-derived respiration. *IEEE Transactions on Biomedical Engineering*, *57*(4), 821–829.
25. Levin, F. K. (1978). The reflection, refraction, and diffraction of waves in media with an elliptical velocity dependence. *Geophysics*, *43*(3), 528–537.
26. Li, S. A., Bao, L., & Chrostowski, M. (2017). Investigating the effects of a personalized, spectrally altered music-based sound therapy on treating tinnitus: a blinded, randomized controlled trial. *Audiology and Neurotology*, *21*(5), 296–304.
27. Liu, B., Zhao, W., & Jiang, Y. (2016). Apparent negative reflection with the gradient acoustic metasurface by integrating supercell periodicity into the generalized law of reflection. *Scientific Reports*, *6*(1), 38314.
28. Lyu, L., Liu, J., Liu, H., Liu, C., Lu, Y., Sun, K., ... & Wujcik, E. K. (2018). An overview of electrically conductive polymer nanocomposites toward electromagnetic interference shielding. *Engineered Science*, *2*(59), 26–42.
29. Miao, J., Chapman, H. N., Kirz, J., Sayre, D., & Hodgson, K. O. (2004). Taking X-ray diffraction to the limit: Macromolecular structures from femtosecond X-ray pulses and diffraction microscopy of cells with synchrotron radiation. *Annu. Rev. Biophys. Biomol. Struct.*, *33*, 157–176.
30. Miller, J. D. (1974). Possible importance of head-diffraction and ear canal resonance for speech perception and hearing-aid design. *The Journal of the Acoustical Society of America*, *55*(2), 462–462.
31. Millikan, R. A. (1923). Coefficients of slip in gases and the law of reflection of molecules from the surfaces of solids and liquids. *Physical review*, *21*(3), 217.
32. Mokdad, R., El-Hafidi, I., & Meyrueis, P. (2004, August). Holographic interferometry deformations metrology by using AR modeling. In *Optical Micro- and Nanometrology in Manufacturing Technology*. SPIE. (Vol. 5458, pp. 136–140).

33. Mulkers, J., Van Waeyenberge, B., & Milošević, M. V. (2018). Tunable Snell's law for spin waves in heterochiral magnetic films. *Physical Review B*, *97*(10), 104422.
34. Ospina, J. F. P., Contreras, V., Estrada-Morales, J., Baresch, D., Ealo, J. L., & Volke-Sepúlveda, K. (2022). Particle-size effect in airborne standing-wave acoustic levitation: Trapping particles at pressure antinodes. *Physical Review Applied*, *18*(3), 034026.
35. Papp, Á., Porod, W., & Csaba, G. (2021). Nanoscale neural network using non-linear spin-wave interference. *Nature Communications*, *12*(1), 6422.
36. Polifke, W., Paschereit, C. O., & Döbbeling, K. (2001). Constructive and destructive interference of acoustic and entropy waves in a premixed combustor with a choked exit. *Int. J. Acoust. Vib*, *6*(3), 135–146.
37. Prévost, R., Coulon, M., Bonacci, D., LeMaitre, J., Millerioux, J. P., & Tourneret, J. Y. (2012, August). Interference mitigation and error correction method for AIS signals received by satellite. In *2012 Proceedings of the 20th European Signal Processing Conference (EUSIPCO)* (Vol. 2, pp. 46–50).
38. Rudenko, O. V. (2009). Nonlinear standing waves, resonance phenomena, and frequency characteristics of distributed systems. *Acoustical Physics*, *55*, 27–54.
39. Saito, N., & Inoue, Y. (2003). Effects of thickness extension mode resonance oscillation of acoustic waves on catalytic and surface properties. IV. Activation of a Ag catalyst for ethanol decomposition by overtone resonance frequencies. *The Journal of Physical Chemistry B*, *107*(9), 2040–2045.
40. Shapira, A., Naor, L., & Arie, A. (2015). Nonlinear optical holograms for spatial and spectral shaping of light waves. *Science Bulletin*, *60*, 1403–1415.
41. Shokri-Ghadikolaei, H., & Fischione, C. (2015). The transitional behavior of interference in millimeter wave networks and its impact on medium access control. *IEEE Transactions on Communications*, *64*(2), 723–740.
42. Siddiqui, M. U. A., Qamar, F., Ahmed, F., Nguyen, Q. N., & Hassan, R. (2021). Interference management in 5G and beyond network: Requirements, challenges and future directions. *IEEE Access*, *9*, 68932–68965.

43. Smallman, L. A., Hill, S. L., & Stockley, R. (1984). Reduction of ciliary beat frequency in vitro by sputum from patients with bronchiectasis: a serine proteinase effect. *Thorax*, *39*(9), 663–667.
44. Suhayda, J. N. (1974). Standing waves on beaches. *Journal of Geophysical Research*, *79*(21), 3065–3071.
45. Tanabe, K., Matsumoto, R., Ohe, J. I., Murakami, S., Moriyama, T., Chiba, D., ... & Ono, T. (2014). Real-time observation of Snell's law for spin waves in thin ferromagnetic films. *Applied Physics Express*, *7*(5), 053001.
46. Tonner, B. P., Han, Z. L., Harp, G. R., & Saldin, D. K. (1991). Scattered-wave integral-transform method of holographic-image reconstruction from forward-scattering diffraction patterns. *Physical Review B*, *43*(18), 14423.
47. Vorob'ev, N. S., Davydov, M. A., Shipilov, K. F., & Shmaonov, T. A. (1981). Formation of beats between Stokes components of stimulated Raman scattering excited in a system comprising two media. *Soviet Journal of Quantum Electronics*, *11*(2), 241.
48. Yeung, R. W., & Sphaier, S. H. (1989). Wave-interference effects on a truncated cylinder in a channel. *Journal of Engineering Mathematics*, *23*(2), 95–117.
49. Zetterstrom, O., Pucci, E., Padilla, P., Wang, L., & Quevedo-Teruel, O. (2019). Low-dispersive leaky-wave antennas for mmWave point-to-point high-throughput communications. *IEEE Transactions on Antennas and Propagation*, *68*(3), 1322–1331.
50. Zhadobov, M., Chahat, N., Sauleau, R., Le Quement, C., & Le Drean, Y. (2011). Millimeter-wave interactions with the human body: State of knowledge and recent advances. *International Journal of Microwave and Wireless Technologies*, *3*(2), 237–247.
51. Zhang, L., Andersen, K. H., Dieckmann, U., & Brännström, Å. (2015). Four types of interference competition and their impacts on the ecology and evolution of size-structured populations and communities. *Journal of Theoretical Biology*, *380*, 280–290.

CHAPTER 6

Beyond Sound and Light: Other Types of Waves

Contents

6.1. INTRODUCTION

The realm of waves encompasses a vast expanse that transcends the limits of human perception, encompassing both visible and audible phenomena. Sound and light waves are widely recognized and have significant impacts on our everyday lives. However, there is a diverse range of waves that extend throughout the electromagnetic spectrum as well as beyond. The current chapter, entitled "Waves: Beyond Sound and Light," explores the captivating realm of waves, which encompass diverse modes of energy propagation and interactions. This study aims to examine the different forms of waves in both the natural world as well as technological applications, ranging from the ubiquitous radio waves that facilitate communication and enjoyment to the formidable seismic waves that contribute to the formation of the Earth's landscape (Santacruz et al., 2020).

This chapter will explore the topic of electromagnetic waves, which includes microwaves, radio waves, infrared radiation, ultraviolet radiation, X-rays, as well as gamma rays. These electromagnetic waves, imperceptible to the human eye, serve as crucial components in the fields of wireless communication, imaging in medicine, and astronomical research. Additionally, we will explore the domain of water waves, observing their elegant oscillations on the surfaces of lakes and oceans, as well as the formidable surges that give rise to devastating tsunamis. However, it should be noted that waves are not solely limited to the observable and tangible phenomena experienced on Earth. In this study, we shall investigate seismic waves, which offer valuable insights into the Earth's interior and the geological forces responsible for shaping its structural composition. Gravitational waves, a remarkable outcome of Einstein's theory of general relativity, cause the expansion and contraction of spacetime during the occurrence of significant cosmic phenomena (Zhang & Northoff, 2022).

Throughout the course of this intriguing chapter, we will explore the manifold attributes, actions, and pragmatic uses of these waves. Through the examination of various types of waves, ranging from the well-known transmission of information through familiar waves to the more intricate disturbances in the structure of spacetime, a more profound comprehension of the role waves play in shaping our knowledge of the universe will be attained. Let us commence this captivating investigation into the realm of waves extending beyond the auditory and visual spectrums (Taniguchi et al., 2014).

6.2. ELECTROMAGNETIC (EM) WAVES

The concept of electromagnetic waves was initially proposed by James Clerk Maxwell before being experimentally verified by Heinrich Hertz. Electromagnetic waves arise from the oscillations occurring between an electric field and a magnetic field. This article aims to provide a comprehensive examination of the definition as well as formation of electromagnetic waves, as well as an in-depth analysis of their graphical and mathematical representations.

Electromagnetic waves, commonly referred to as EM waves, are a fundamental concept in physics. Electromagnetic radiation is comprised of electromagnetic waves, which are generated through the interaction between an electric field and a magnetic field. Electromagnetic waves can be characterized as the amalgamation of magnetic and electric fields that oscillate. Electromagnetic waves are regarded as solutions to Maxwell's equations, which serve as the foundational equations of electrodynamics (Hikita et al., 2007).

6.2.1. Formation of Electromagnetic Waves

The following are the steps in formation of EM waves:

- In general, the presence of an electric charge generates an electric field. The electric field exerts a force on particles that are not charged. Positive charges experience acceleration in the same direction as the electric field, while negative charges experience acceleration in the opposite direction to the electric field.
- The generation of a magnetic field is a result of the motion of a charged particle. The magnetic field exerts a force on other particles in motion. The force exerted on these charges consistently acts at a right angle to the axis of their velocity, resulting in a change in orientation of the velocity but not the magnitude of the velocity (Sun et al., 2020).
- The production of the electromagnetic field is a result of the acceleration of a charged particle. Electromagnetic waves consist of magnetic and electrical fields that propagate through a vacuum at the constant speed of light, denoted as c. An accelerating particle with charge refers to a scenario in which the charged particle undergoes oscillatory motion around a stable equilibrium position. If the charged particle oscillates at a frequency denoted

as f, it generates an electromagnetic wave with the same frequency f. The wavelength of the wave is determined by the equation $\lambda = c/f$, where λ represents the wavelength, c represents the speed of the wave, and f represents the frequency of the wave. Electromagnetic waves facilitate the transmission of energy across the expanse of space (Hui et al., 2023).

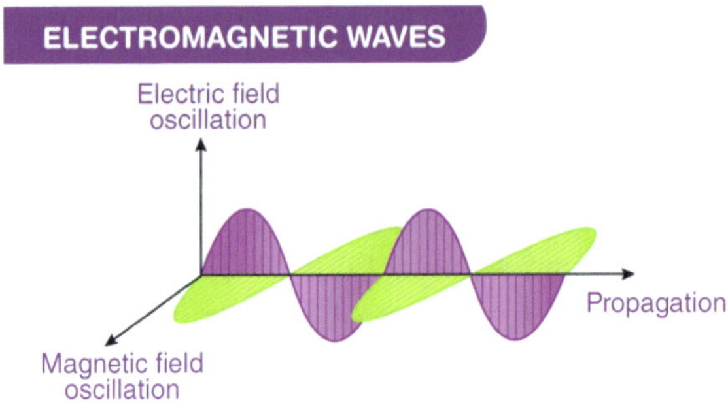

Figure 6.1. Representation of Electromagnetic Waves (Source: BYJU's, Creative Commons License).

Electromagnetic waves are represented by a sinusoidal waveform (Figure 6.1). The phenomenon under consideration comprises of magnetic and electric fields that exhibit temporal variations. These fields are mutually perpendicular and additionally perpendicular to the axis in which the waves propagate. Electromagnetic waves possess a transverse characteristic. The apogee of the wave is commonly referred to as the crest, whereas the nadir is designated as the trough. In a vacuum, electromagnetic waves propagate at a consistent velocity of 3 x 10^8 meters per second (Yuan et al., 2021).

6.2.2. Mathematical Representation

The plane electromagnetic wave propagating in the x-direction can be expressed as:

$$(x,t) = coscos(kx - Wt + \Phi)$$

$$(x,t) = coscos(kx - Wt + \Phi)$$

Maxwell proposed the fundamental concept of electromagnetic radiation, while Hertz empirically validated the presence of electromagnetic waves.

The vector cross product of the field of electricity as well as magnetic field determines the direction of transmission of the electromagnetic wave. The information is provided as follows (Ambrose et al., 1999):

$$E^{\rightarrow} \times B^{\rightarrow}$$

6.2.3. Electromagnetic Wave Equation

- The transmission of electromagnetic waves in a vacuum or across a medium is described by the electromagnetic wave equation.
- The equation governing electromagnetic waves is the second-order partial differential equation.
- The given equation represents a three-dimensional manifestation of the wave equation.
- The equation can be expressed in its homogeneous form (Masood et al., 2010).

$$\left(\upsilon_{ph}^{2}\nabla^{2}-\frac{\partial^{2}}{\partial t^{2}}\right)E=0$$

$$(\upsilon_{ph}^{2}\nabla^{2}-\frac{\partial^{2}}{\partial t^{2}})B=0$$

where,

$$\upsilon_{ph}=\frac{1}{\sqrt{\mu\epsilon}}$$

6.2.4. Applications of Electromagnetic Waves

There exist several applications of electromagnetic waves.

- Electromagnetic radiation has the ability to transfer energy without the need for a medium or the presence of any material substance.
- Electromagnetic waves are of significant importance in the field of communication technology.
- Electromagnetic waves are utilized in the operation of RADARS.
- Ultraviolet (UV) rays are employed for the purpose of identifying counterfeit banknotes. Authentic banknotes do not exhibit fluorescence when exposed to ultraviolet (UV) light.

- Infrared radiation finds application in night vision technology and is also utilized in the operation of security cameras (Chen et al., 2018).

6.3. WATER WAVES

When positioned on a shoreline, it is customary to observe the physical movement of water towards the land, characterized by the rolling motion of water waves (Figure 6.2). Waves are generated through the interaction of wind with a stationary body of water along the shoreline, resulting in perturbations that propagate in a circular trajectory. Waves generated by water are a type of surface wave that exhibit an amalgam of longitudinal as well as transverse wave characteristics. The distortions propagate at a rate equal to the speed of the wave, whereas the water molecules stay stationary in their respective positions. The majority of oceanic waves are generated as a result of wind activity, with the waves propagating towards the coastal regions facilitating the transfer of wind energy from offshore locations. Nevertheless, the energy is directed towards the shoreline (Longuet-Higgins, 1953).

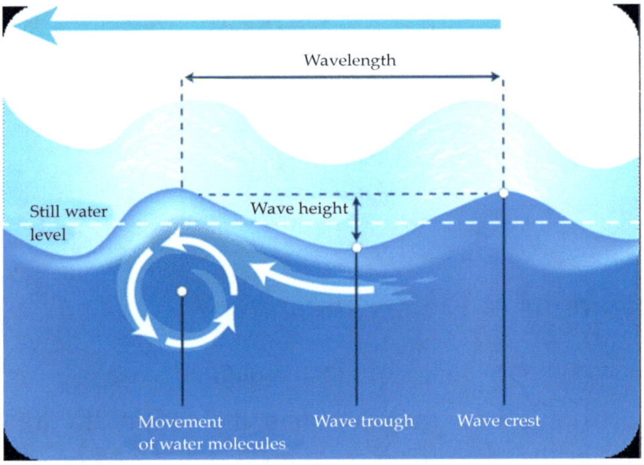

Figure 6.2. Ocean movements and waves (Source: CK-12, Creative Commons License).

6.3.1. Concept of Water Wave Tsunami

A tsunami is a natural event characterized by a series of powerful oceanic waves that result in a significant increase in water level. A tsunami is a catastrophic oceanic phenomenon that typically arises from a subaquatic seismic event, coastal mass movement, or volcanic discharge. Tsunamis, commonly referred to as tidal waves, are not associated with tidal activity. The term "tsunami" originates from the Japanese language, specifically denoting a "harbor wave" (Figure 6.3) (Ramadan et al., 2011).

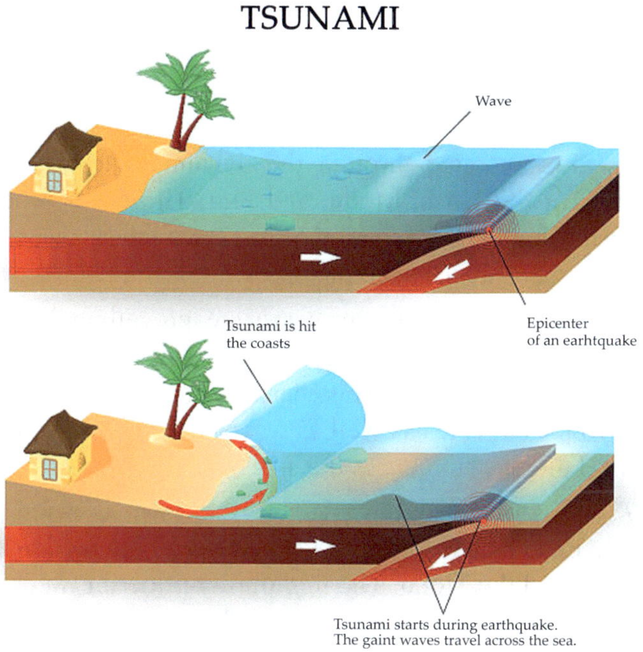

Figure 6.3. Illustration of the concept of Water Tsunami (Source: TTSZ, Creative Commons License).

The aforementioned waves emanate outward, resulting in the generation of impulses with velocities of up to 500 miles (800 km) per hour, and attaining peak elevations of 100 feet (30 meters) in proximity to coastal regions. Tsunamis are primarily triggered by seismic activity occurring along convergent tectonic plate boundaries. According to the Global Historical Tsunami Database, it has been observed that earthquakes have accounted for more than 80% of the anticipated tsunamis since the year 1900 (Sahoo & Chakraverty, 2022).

Tip

Tsunamis are a specific type of water waves caused by large disturbances, such as underwater earthquakes or volcanic eruptions. Understanding the physics behind tsunami formation is crucial in predicting and preparing for potential tsunamis to mitigate their impact on coastal areas.

6.3.2. Water Wave Formation

Water waves are generated through the dynamic interplay between the celestial bodies of the Earth, Moon, and Sun. The development of water waves can be attributed to the wind that traverses the surfaces of bodies of water, such as rivers and oceans (Annenkov & Shrira, 2006).

- ***Wind waves***: Wind waves are the prevailing form of waves, characterized by the transfer of substantial energy from the wind to the water's surface. This energy exchange leads to the formation of ripples in lakes as well as waves in oceans. The magnitude of wind directly influences the size of waves.
- ***Tidal waves***: Tidal waves, also known as wind waves, are generated as a result of the gravitational interaction between the Earth and the Moon. The immobility of ocean water is attributed to the gravitational force exerted by the Earth. However, the gravitational pull exerted by the moon on the planet induces oscillatory motion in the ocean water, resulting in the formation of waves that exhibit vertical undulations. This phenomenon arises when the moon is in closer proximity to a specific side of the Earth. The phenomenon referred to as high tide occurs in one part of the Earth, while the opposite side, in which the moon is not directly aligned, experiences low tide (Jian & Changlong, 2006).
- ***Underwater explosions***: This phenomenon constitutes an additional factor contributing to the formation of waves. Enormous oceanic waves, such as tsunamis, are generated as a consequence of seismic activity occurring beneath the water's surface. Additionally, underwater explosions caused by volcanoes, landslides, as well as meteors can lead to the formation of immense and devastating waves (Brett & Krelle, 2018).

There exist three primary categories of water waves, namely seismic sea waves (tsunamis), swell waves, as well as wind surges.

6.3.3. Water Wave Characteristics

The features of water waves are as follows:

- *Wave crest and trough*: The highest and lowest points in a wave are referred to as the crest as well as trough, respectively.
- *Wave amplitude*: The term "wave amplitude" refers to half of the wave height.
- *Wave height*: Wave height refers to the vertical measurement from the highest point of a wave crest to the lowest point of a wave trough.
- The wave period refers to the temporal interval between successive wave troughs and crests as they traverse a stationary reference point.
- *Wavelength*: The horizontal separation between two consecutive peaks.
- *Wave speed*: Wave speed, denoted as the rate by which a wave propagates through a medium, particularly water, is commonly quantified in units of knots.
- *Wave frequency*: Within a one-second time interval, the total quantity of waves that traverse a specific point (Wu et al., 2020).

6.4. RADIO WAVES

Radio waves are a form of electromagnetic radiation characterized by their extended wavelengths within the electromagnetic spectrum, commonly exhibiting frequencies that are typically 300 gigahertz (GHz) and lower (Figure 6.4). The inclusion of a citation is necessary to support the validity and reliability of the information provided. At a frequency of 300 GHz, the wavelength that corresponds measures 1mm, which is of lesser magnitude compared to the diameter of a single grain of rice. When the frequency is 30 Hz, the wavelength is approximately 10,000 kilometers (6,200 miles) greater than the Earth's radius. The wavelength of a radio wave exhibits an inverse relationship with its frequency due to the constancy of its velocity. Similar to other electromagnetic waves, the radio waves propagate at the velocity of light in a vacuum, while in the Earth's atmosphere, their speed is marginally reduced. Radio waves are produced when charged particles experience acceleration, such as electric currents that vary over time. Naturally arising radio waves are released by both atmospheric phenomena such as lightning and celestial entities, and are integral components of the blackbody radiation that is emitted by objects with elevated temperatures (Van de Hulst, 1945).

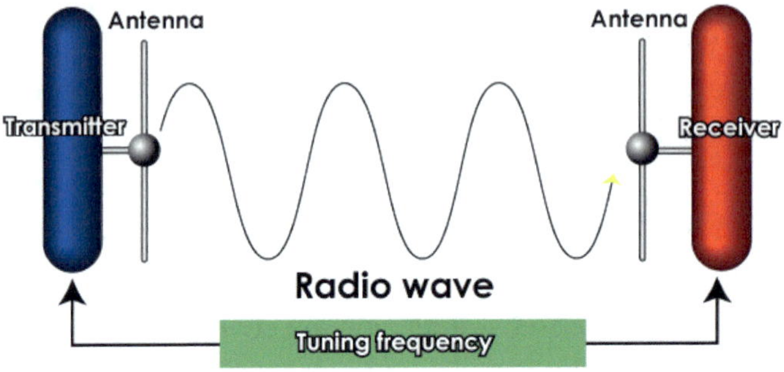

Figure 6.4. Propagation of radio waves (Source: Our Education, Creative Commons License).

Artificial generation of radio waves is accomplished through the utilization of an electronic apparatus known as a transmitter that is interconnected with an antenna that facilitates the emission of said waves. The signal is received by a separate antenna that is linked to a radio receiver, which then proceeds to process the received signal. Radio waves are extensively utilized in contemporary technology for various purposes such as fixed as well as mobile radio communication, transmission, radar as well as radio navigation systems, satellites for communication, wireless computer networks, and numerous other applications. The Earth's atmosphere exhibits varying propagation characteristics for radio waves of different frequencies. Longer waves, for instance, possess the ability to diffract around physical barriers such as mountains and conform to the Earth's surface (known as ground waves). Conversely, shorter waves can reflect off the ionosphere and return to the Earth's surface beyond the observable horizon (known as skywaves). Lastly, significantly shorter wavelengths experience minimal bending or diffraction, resulting in their propagation being limited to the line of sight and restricted to distances within the visual horizon (Jansky, 1933).

In order to mitigate potential disruptions among various users, the utilization and production of radio waves are subject to stringent legal regulations. These regulations are overseen by the International Telecommunication Union (ITU), an international organization responsible for coordinating these efforts. The ITU defines radio waves as "electromagnetic waves with frequencies below 3,000 GHz that propagate through space without the aid of artificial guidance." The radio spectrum is partitioned into various radio bands based on frequency, which are assigned for different purposes (Brown, 1984).

6.5. SEISMIC WAVES

A seismic wave refers to a propagating disturbance of acoustic energy that traverses the Earth or any other celestial body. Seismic activity can arise from various natural phenomena such as earthquakes, volcanic eruptions, magma displacement, significant landslides, as well as from large-scale artificial explosions that generate low-frequency acoustic energy. Seismologists investigate seismic waves through the utilization of seismometers, hydrophones (when studying waves in water), or accelerometers, which enable the recording of these waves. Seismic waves can be differentiated from seismic noise, also known as ambient vibration, as the latter refers to continuous, low-intensity vibrations that originate from a range of natural and human-induced sources. The velocity at which a seismic wave propagates is contingent upon the characteristics of the medium, including its density and elasticity, as well as the specific type of wave being considered. The velocity of seismic waves generally exhibits an upward trend as depth increases within the Earth's crust as well as mantle, yet experiences a significant decrease upon transitioning from the mantle to the outer core of the Earth (Hennino et al., 2001).

Earthquakes give rise to various types of waves characterized by distinct velocities. The seismic observatory records the varying travel times of seismic waves, which aids scientists in determining the hypocenter of an earthquake. The utilization of seismic wave refraction or reflection in geophysics serves as a means to investigate the internal composition of the Earth. Researchers occasionally generate and quantify vibrations in order to investigate shallow subsurface structures (Honda, 1962).

Within the realm of seismic waves, a fundamental categorization can be made among body waves, which propagate through the Earth's interior, and surface waves, that propagate along the Earth's outermost layer.

There are additional modes of wave transmission that are not discussed in this article. While these modes are of lesser significance for waves on Earth, they hold importance in the field of asteroseismology.

- Body waves propagate within the Earth's interior.
- Surface waves propagate along the surface. Surface waves exhibit a slower decay rate with increasing distance compared to body

waves, as the latter propagate in three-dimensional space.

- The amplitude of particle motion associated with surface waves is greater compared to that of body waves, resulting in a higher potential for causing damage (Elkhoury et al., 2006).

6.6. MICROWAVES

Microwaves, along with radio waves, ultraviolet radiation, X-rays, and gamma-rays, belong to the category of electromagnetic radiation. Microwaves possess a diverse array of applications, encompassing various fields such as telecommunications, radar technology, and notably, culinary practices (Justesen, 1975).

Electromagnetic radiation is propagated through the medium of waves or discrete particles, exhibiting variations in both wavelength and frequency. The electromagnetic spectrum (EM spectrum) encompasses a wide range of wavelengths. The electromagnetic spectrum is commonly categorized into seven distinct regions based on a progressive decrease in wavelength and an increase in energy and frequency. The commonly recognized classifications encompass radio frequencies, microwaves, infrared (IR), light that is visible, ultraviolet (UV), X-rays, and gamma-rays. Microwaves are situated within the electromagnetic spectrum, occupying a position between radio waves and infrared light (Rosen et al., 2002).

6.6.1. Communications and Radar

According to the Federal Communications Commission (FCC), microwaves are predominantly employed in point-to-point communication systems for the transmission of various forms of information, encompassing voice, data, and video, in both analog and digital configurations. In addition, they find application in the realm of SCADA (supervisory control and data acquisition) systems, where they are utilized for the purpose of remotely monitoring and controlling machinery, switches, valves, and signals (Sobol & Tomiyasu, 2002).

Radar is a significant application of microwaves. The term "radar" originated as an abbreviation for Radio Detection and Ranging. In the pre-World War II era, British radio engineers made a significant discovery pertaining to the behavior of short-wavelength radio waves. They observed that these waves possessed the ability to be reflected off remote entities such as ships and aircraft. By employing exceptionally sensitive directional antennas, these engineers were able to detect the resultant signals, thereby enabling the determination of the presence and precise locations of said objects. The term "radar" has attained such widespread usage that it has acquired lexical status as a standalone word, encompassing systems that employ either microwaves or radio waves (Donnet & Longstaff, 2006).

An obscure historical detail pertains to the construction of an initial radar installation situated at the summit of Kahuku Point, located on the northernmost extremity of Oahu. Based on the official website of the state of Hawaii, it is documented that the station successfully detected the initial wave of Japanese aircraft en route to launch an attack on Pearl Harbor at a distance of 132 miles (212 kilometers). Nevertheless, due to the system's limited operational period of merely a couple of weeks, it was deemed to lack reliability, resulting in the disregard of the warning. Throughout the duration of the conflict, radar underwent significant enhancements and refinements, subsequently establishing itself as a fundamental component of both national defense strategies as well as civilian air-traffic control systems (Meinel, 1995).

The utilization of the Doppler Effect has facilitated the discovery of numerous additional applications for radar. One instance illustrating the Doppler Effect may be observed through the presence of an approaching ambulance. As the siren approaches, its auditory frequency appears to increase, culminating in a wailing sound as it passes by. Subsequently, as the siren moves away, it appears to exhibit a decrease in frequency (Baylis et al., 2014).

According to Robert Mayanovic, a physics professor at Missouri State University, Doppler radar, frequently utilizing microwaves, is employed in the domains of air-traffic control as well as vehicular speed-limit enforcement. If an object is in close proximity to the antenna, the microwaves that are reflected back become compressed, resulting in a decrease in wavelength and an increase in frequency. In contrast, it can be observed that waves reflected from objects in motion away from the observer exhibit elongation, resulting in longer wavelengths and a lower frequency. The determination of an object's speed toward or away from the antenna may be achieved by measuring the frequency shift (Stockman, 1948).

This principle finds widespread application in various domains, such as the development of basic motion detection systems, the utilization of radar guns for enforcing speed limits, the implementation of radar altimeters, and the utilization of weather radar to track the three-dimensional movement of water droplets within the atmosphere. The aforementioned applications are commonly referred to as active sensing, as they involve the transmission of microwaves and subsequent analysis of the received and reflected signals. Passive sensing involves the observation and analysis of microwaves emitted by natural sources. A significant portion of these observations is carried out by satellites, which are employed to observe either the Earth in a retrospective manner or the outer space (Li et al., 2017).

6.6.2. Microwave Heat Sources

One of the prevailing applications of microwaves is expediting the process of heating food. The feasibility of microwave ovens arises from the utilization of microwaves for the transmission of thermal energy. The fortuitous nature of this phenomenon's discovery was entirely unintentional. In the book titled "They All Laughed...: From the Light Bulbs to Lasers: The Fascinating Stories Behind the Great Inventions That Have Changed Our Lives" (HarperCollins, 1992), Ira Flatow, the author, provides an account of the invention of the microwave oven. The narrative revolves around Percy L. Spencer, a distinguished figure in the field of electronics and a decorated war veteran, who, in the aftermath of World War II, was conducting a tour of one of the laboratories at the Raytheon Company. Spencer came to a halt in the presence of a magnetron, which serves as the power tube responsible for propelling a radar system. Abruptly, he became aware of the fact that a confectionery item stored within his pocket had commenced the process of liquefaction. Subsequent inquiry prompted him to produce the initial quantity of microwave popcorn, alongside the inaugural occurrence of an exploding egg (Lagos et al., 1995). Initially, the early models of microwave ovens were characterized by their substantial size and high cost. However, over time, these appliances have undergone a significant transformation, resulting in a notable decrease in price. As a consequence, microwave ovens have become increasingly accessible to the general public, to the extent that they are now prevalent in households across the globe. Microwave heating systems find application in various industrial sectors, such as food processing, chemical processing, and materials processing, in both batch as well as continuous operations (Constant et al., 1996).

6.6.3. Natural Microwave Sources

Observations in the microwave region are carried out by radio astronomers, who often resort to utilizing high-altitude balloons and satellites due to the atmospheric attenuation that occurs. Nevertheless, arguably the most renowned instance of detecting extraterrestrial microwaves was carried out by a pair of scientists from Bell Labs who were engaged in the development of a telecommunications system utilizing a substantial terrestrial horn antenna. As per the information provided on the NASA Science website, Arno Penzias as well as Robert Wilson, scientists affiliated with Bell Labs, made an accidental but remarkable discovery in 1965. Utilizing long-wavelength microwaves in the L-band range, they successfully detected background noise through the employment of a specialized low-noise antenna. The peculiarity of the auditory phenomenon lied in its omni-directional origin and consistent lack of significant variation in intensity. If the source of this static were terrestrial, like radio transmissions originating from an airport nearby control tower, it would display directional characteristics, emanating from a specific location rather than being uniformly distributed. The scientists at Bell Laboratories quickly recognized that they had fortuitously stumbled upon the cosmic microwave background radiation radiation. The presence of this radiation, permeating the entirety of the cosmos, serves as an indication of its origin, commonly referred to as the Big Bang (Korovin et al., 2004).

The 1978 Nobel Prize in physics was awarded to Penzias and Wilson in recognition of their discovery. Satellites have effectively mapped the cosmic microwave background radiation with a high degree of precision. The aforementioned observations have provided insights into the subtle fluctuations in temperature that ultimately gave rise to the formation of the galactic clusters that are presently observable.

The examination of this background radiation has provided astronomers with valuable insights into the composition of the universe. It is currently believed that approximately 95 percent of the cosmos consists of energy and matter that cannot be detected using conventional instruments. Consequently, these elusive components have been termed dark matter and dark energy. According to certain cosmic models, future examination of this background radiation has the potential to provide additional insights into the events that transpired in the early stages of the universe's formation, and conceivably, even prior to the existence of our current universe (Thurber et al., 2010).

6.7. X-RAYS

X-rays are a form of electromagnetic radiation characterized by their exceptionally short wavelengths and high frequencies (Figure 6.5). These wavelengths typically span from approximately 10−8 to 10−12 meters, while the corresponding frequencies range from about 1016 to 1020 hertz (Hz).

X-rays are frequently generated through the acceleration or deceleration of charged particles. Illustrative instances encompass the collision of an electron beam with a metal plate within an X-ray tube, as well as the circulation of an electron beam within a synchrotron particle accelerator and storage ring. Furthermore, it is worth noting that atoms in a state of high excitation possess the ability to release X-rays that exhibit distinct wavelengths, which are directly related to the energy level gaps within the atoms. The X-ray area of the electromagnetic spectrum is situated significantly beyond the range of wavelengths that are perceptible to the human eye. Nevertheless, the transmission of X-rays through various substances, such as biological tissue, may be documented by means of photographic films along with various types of detectors. The examination and interpretation of X-ray images of the human body constitute a highly valuable modality in medical diagnostics.

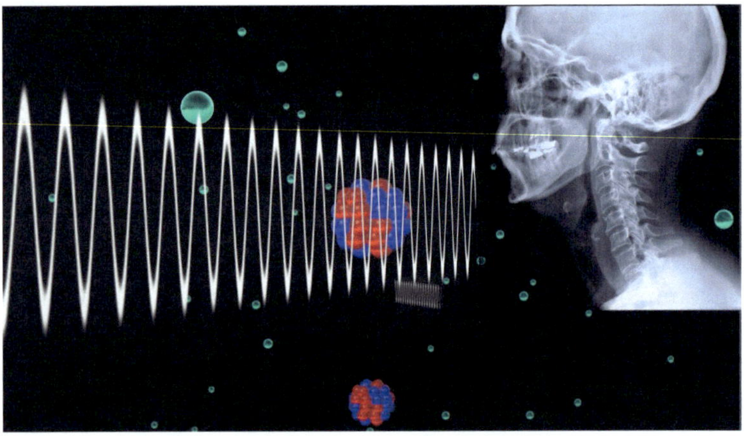

Figure 6.5. Schematic of the X-rays (Source: NIBIB gov, Creative Commons License).

X-rays are classified as a type of ionizing radiation due to their ability to induce the ejection of electrons from neutral atoms when they interact with matter. The energy of the X-rays is transferred to the matter through the process of ionization. X-rays have the potential to induce detrimental biochemical alterations in chromosomes, genes, and various cellular constituents when traversing living tissue. The investigation of the biological consequences of ionizing radiation remains an ongoing area of research, characterized by intricate dynamics and a strong reliance on the duration and magnitude of exposure (refer to radiation injury). X-ray radiation therapies utilize these phenomena to counteract the proliferation of malignant tumors (Hall, 1936).

The discovery of X-rays took place in 1895, when German physicist Wilhelm Konrad Röntgen conducted experiments to investigate the impact of electron beams, known as cathode rays at the time, on electrical discharges occurring in low-pressure gases. Röntgen made a significant discovery, which involved observing that a screen covered with a fluorescent substance exhibited luminescence regardless of it was protected from the direct visible as well as ultraviolet radiation emitted by the discharge of a gaseous tube. The individual inferred that an imperceptible form of radiation emitted from the tube traversed the surrounding atmosphere, resulting in the fluorescence of the screen. Röntgen successfully demonstrated that the radiation accountable for the phenomenon of fluorescence originated from the precise location where the electron beam made contact with the glass wall of the discharge tube. The transparency of opaque objects positioned between the tube and the screen was observed in relation to the newly discovered radiation. Röntgen effectively showcased this phenomenon by capturing a photographic representation of the skeletal structure of a human hand. The global scientific and popular community responded with great enthusiasm to his groundbreaking identification of Röntgen rays, commonly known as X-rays. This significant achievement, combined with the subsequent discoveries of radioactivity in 1896 and the electron in 1897, marked the advent of atomic research and the commencement of the modern physics era (Kane et al., 1986).

6.7.1. Fundamental characteristics

6.7.1.1. Wave Nature

X-rays are classified as a type of electromagnetic radiation, sharing fundamental physical characteristics with other well-known parts of the electromagnetic spectrum, such as light that is visible, infrared radiation, as well as ultraviolet radiation. Similar to other types of electromagnetic radiation, X-rays may be characterized as interconnected waves of magnetic and electric fields that propagate at the velocity of light, approximately 300,000 kilometers (or 186,000 miles) per second. The demonstration and measurement of the characteristic wavelengths as well as frequencies of waves can be achieved by observing the effects of interference that occur when two or more waves overlap in space. X-rays also possess characteristics akin to particles, as they can be characterized as a stream of photons that bear discrete quantities of momentum and energy. The duality exhibited by all forms of matter and radiation is a fundamental characteristic that is thoroughly expounded upon in the theory of quantum mechanics.

Remember

X-rays have a high penetrating power and can pass through soft tissues, making them invaluable in medical imaging, such as X-ray radiography, CT scans, and fluoroscopy, to visualize the internal structures of the human body.

After Röntgen's discovery, there was an immediate suspicion that X-rays could be classified as a type of electromagnetic radiation. However, confirming this hypothesis proved to be a challenging task. X-rays are characterized by their significantly reduced wavelengths, typically measuring 1,000 times shorter compared to the wavelengths of visible light. Due to these circumstances, as well as the inherent challenges associated with generating and detecting the novel type of radiation, the elucidation of X-rays' characteristics occurred progressively during the initial years of the 20th century (Marshall & Santos, 1988).

The wave nature of X-rays was initially demonstrated by Charles Glover Barkla, a British physicist, in the year 1906. Barkla's experiment involved the observation of X-rays being "polarized" through their scattering from a solid material. The concept of polarization pertains to the alignment of oscillations in a transverse wave. It is important to note that all electromagnetic waves can be described as transverse oscillations involving electric and magnetic fields. The establishment of the remarkably short wavelengths of X-rays, initially suggested through diffraction experiments involving the passage of these rays through narrow slits, was definitively confirmed in 1912 by the groundbreaking research conducted by Max von Laue, a German physicist, along with his students Walter Friedrich as well as Paul Knipping. Laue proposed that the orderly configurations of atoms within crystals possess the inherent capability to function as innate three-dimensional diffraction gratings. The atomic spacings commonly observed in crystals are typically on the order of 1 angstrom, equivalent to 1×10^{-10} metre. This particular spacing is well-suited for generating diffraction phenomena when interacting with electromagnetic radiation of similar wavelengths. The veracity of Laue's predictions was confirmed by Friedrich and Knipping through the utilization of photographic techniques to capture diffraction patterns resulting from the transmission of X-rays via a crystal composed of zinc sulfide. The conducted experiments provided evidence that X-rays possess wavelengths approximately equal to 1 angstrom, thereby validating the hypothesis that atoms within crystals exhibit a consistent and organized arrangement (Adamovich et al., 1984).

In the subsequent year, the British physicist William Lawrence Bragg formulated a notably straightforward framework for elucidating the phenomenon of X-ray scattering occurring within the parallel atomic layers of a crystal. The Bragg law elucidates the correlation between the optimal diffraction angles of X-rays from a crystal, the wavelength of the X-rays, and the interatomic spacing within the crystal lattice. The design of the inaugural X-ray spectrometer was formulated by William Henry Bragg, a physicist, who drew inspiration from the analysis conducted by his son, Bragg. The duo employed an X-ray spectrometer to conduct significant investigations on the wavelength distribution in X-ray beams as well as the crystal structures of numerous prevalent solids. This notable accomplishment led to their joint receipt of the Nobel Prize for Physics in 1915 (Cox et al., 2007).

6.7.1.2. Particle Nature

During the initial years of the 1920s, investigations involving the dispersion of X-rays from solids through experimental methods played a pivotal role in the establishment of the particle-like characteristics inherent in electromagnetic radiation. In 1905, Albert Einstein, a physicist from Germany, put forth the proposition that electromagnetic radiation is composed of discrete units known as quanta, which were later named photons. Each photon possesses an energy value of hf, where h represents Planck's constant (approximately 6.6 × 10−34 joule·second) and f denotes the frequency of the radiation. The hypothesis proposed by Einstein received substantial support in subsequent investigations of the photoelectric effect, as well as through the achievements of Niels Bohr, a Danish physicist, in developing a model for the hydrogen atom that explained its distinctive both absorption and emission spectra (refer to the Bohr atomic model). Additional confirmation was obtained in 1922 when Arthur Compton, an American physicist, effectively analyzed the phenomenon of X-ray scattering from atoms in a solid as a series of interactions involving collisions among X-ray photons and the weakly bound outer electrons of the atoms.

By applying the principles of conservation of energy and conservation of momentum, Compton derived a description for the wavelength shift of scattered X-rays as an estimate of their scattering angle. This derivation involved adapting the relationship between energy and momentum for a conventional electromagnetic wave to an individual photon. The phenomenon commonly referred to as the Compton Effect involves the transfer of energy and momentum from a photon to an electron through collision, resulting in the recoil of the electron. Consequently, the energy and momentum of the scattered photon are reduced compared to those of the incident photon, leading to the production of X-rays with slightly lower frequency as well as longer wavelength. Compton's meticulous measurements of this minute phenomenon, in conjunction with his proficient theoretical analysis (which was independently formulated by the Dutch scientist Peter Debye), presented compelling substantiation for the presence of photons. The X-ray segment of the electromagnetic spectrum spans a wavelength range of approximately 10^{-8} to 10^{-12} meters, which corresponds to an array of photon energies from approximately 100 electron volts (eV) to 1 million electron volts (MeV) (Henriksen et al., 2018).

6.7.2. Applications

The distinctive attributes of X-rays, including their capacity to penetrate materials that are not transparent to visible light, their atomic-scale wavelengths, and the significant energy carried by each X-ray photon, contribute to a diverse array of applications in industries, medicine, and scientific research. Various specialized X-ray sources of information, detectors, and evaluation techniques have been specifically designed to investigate a wide array of inquiries, spanning from the examination of the interactions involving basic molecules to the elucidation of the intricate structure of the human brain (Bergé, 1998).

X-ray imaging of the human body is an essential diagnostic modality in contemporary medical practice. Medical imaging techniques enable the noninvasive identification of dental caries, skeletal fractures, foreign bodies, and pathological states, including cancer. X-ray images are capable of effectively distinguishing among bone and soft tissue. To enhance the differentiation of various soft tissue regions, a contrast medium can be administered. This medium, which can be in the form of a liquid or gas, exhibits a higher level of opacity to X-rays compared to the surrounding tissues (refer to diagnostic imaging). During the 1970s, a novel and influential imaging technique known as computed tomography (CT) emerged. Currently, CT scans are extensively utilized to generate intricate and high-resolution cross-sectional images of internal structures and organs. These scans exhibit a significantly higher sensitivity to small density variations compared to conventional X-ray images (Buttgen et al., 2004).

Similar to other types of ionizing radiation, X-rays induce biochemical alterations in living cells. The energy of a high-energy X-ray photon is released through the process of ionization, where electrons are freed from their respective atomic and molecular structures. These unbound electrons have the potential to induce ionization in other neutral species. During this particular process, the formation of reactive ions as well as free radicals occurs, subsequently initiating additional chemical reactions. The radiation-induced chemistry that ensues can disrupt the molecular bonds essential for cellular proliferation and give rise to genetic harm (refer to radiation injury). Although there exist notable health hazards linked to X-ray exposure, radiation therapies utilize these aforementioned effects to address cancerous tumors and hematological disorders like leukemia. X-ray radiation, along with gamma rays of higher energy, is deliberately focused on specific tissues in order to induce molecular damage, which effectively inhibits the proliferation of afflicted cells. Adjacent healthy cells, which are also subjected to ionizing X-rays, generally exhibit a greater capacity for repair. In agricultural industries, the utilization of X-rays and gamma rays for the irradiation of certain foods is employed as a means to selectively impede the proliferation of bacteria, as observed in the context of food preservation (Havelock, 1908).

X-ray imaging serves as a robust diagnostic instrument that facilitates the elucidation of the internal structure and elemental composition of various materials. The significant practicality of X-ray images stems from the distinct absorption characteristics of X-rays based on the varying density, composition, as well as homogeneity of materials. X-rays are commonly employed in the context of airline baggage inspection to facilitate expedient examination of the contents. X-ray imaging is employed in industrial settings to nondestructively identify defects in castings that are not readily observable through direct means. X-ray microscopes possess the ability to amplify X-ray absorption images, enabling the discernment of minute details on a scale as diminutive as approximately 40 nanometers (nm), equivalent to approximately 400 atomic diameters. The resolution achieved in X-ray microscopy is approximately five times higher compared to the resolution achieved by the most advanced visible light microscopes. This enhanced resolution is attributed to the minimal diffraction effects that occur due to the extremely short wavelengths of X-rays. X-ray microscopes commonly utilize X-rays with "soft" characteristics, characterized by wavelengths ranging from 1 to 10 nanometers. These microscopes rely on reflective optics or "zone plates" to achieve the necessary focusing capabilities. For further information on X-ray optics, refer to the field of spectroscopy, while the field of optics provides insights into filtering techniques using zone plates. The transparency of water in the soft X-ray region makes these microscopes well-suited for the examination of biological materials in a water-based setting. An additional advanced method for absorption, known as EXAFS (extended X-ray absorption fine structure), exhibits the capability to discern the arrangement of atoms and molecules at a short range within unstable samples of crystals as well as amorphous solids (Kiselev, 2007).

X-ray diffraction techniques, also known as X-ray crystallography, facilitate the elucidation of crystal structures in a wide range of materials, including inorganic, organic, as well as biological substances. The elucidation of the intricate atomic arrangement of the double-helix polymer known as deoxyribonucleic acid (DNA) gained significant renown through the pioneering efforts of James Watson and Francis Crick, who employed X-ray crystallography techniques pioneered by Maurice Wilkins. X-ray fluorescence (XRF) serves as a supplementary technique for the quantitative assessment of material composition. In this methodology, a specimen is subjected to either an electron beam or a primary X-ray beam, inducing atomic excitations that subsequently generate X-ray emissions exhibiting wavelengths that are distinctive to the constituent elements present in the specimen. The electron microprobe employs a methodology to ascertain the components of sample regions with dimensions on the order of micrometers. X-ray fluorescence as well as diffraction techniques are highly advantageous methodologies employed in the nondestructive examination of art objects. X-ray analysis can be employed to unveil brushstroke techniques, the arrangement of painted-over pigments in oil paintings, the existence of coatings as well as varnishes, as well as the formulations of glasses, porcelain, and enamels (Armstrong et al., 1962).

The aforementioned techniques are significantly improved by the remarkably high levels of X-ray intensities generated in contemporary synchrotron light facilities. Highly intense, brief X-ray pulses, specifically tailored to specific wavelength ranges, are employed for the purpose of investigating chemical reactions occurring on surfaces, examining the electronic configurations of semiconductors as well as magnetic substances, as well as analyzing the composition and operation of proteins and other biological macromolecules. The X-ray laser represents a potentially auspicious avenue for generating high-intensity X-rays. In laboratory settings, the production of coherent X-rays, which serve as a signature of lasing, has been achieved at the longer-wavelength range within the spectral region. In 2009, the Linac Coherent Light Source facility in Menlo Park, California successfully demonstrated lasing at a wavelength of 0.15 nm. However, the construction of a viable device operating at such short wavelengths continued to pose significant technological obstacles (Dacey, 1962).

6.8. GAMMA RAYS

Gamma rays are a form of electromagnetic radiation characterized by their exceptionally short wavelength and high energy levels (Figure 6.6). Gamma rays are generated through the process of radioactive atomic nuclei disintegration and the decay of specific subatomic particles. The gamma-ray as well as X-ray regions of the electromagnetic spectrum exhibit some degree of wavelength overlap according to widely accepted definitions. Specifically, gamma-ray radiation is characterized by wavelengths typically smaller than a few tenths of an angstrom (10–10 meter), while gamma-ray photons possess energies exceeding tens of thousands of electron volts (eV). Theoretical upper limits to the energies of gamma-ray photons as well as lower limits to gamma-ray wavelengths have not been established. Currently, observed energies of gamma-ray photons reach up to several trillion electron volts. These exceptionally high-energy photons are generated in astronomical sources, although the specific mechanisms responsible for their production remain unidentified (Ramaty et al., 1975).

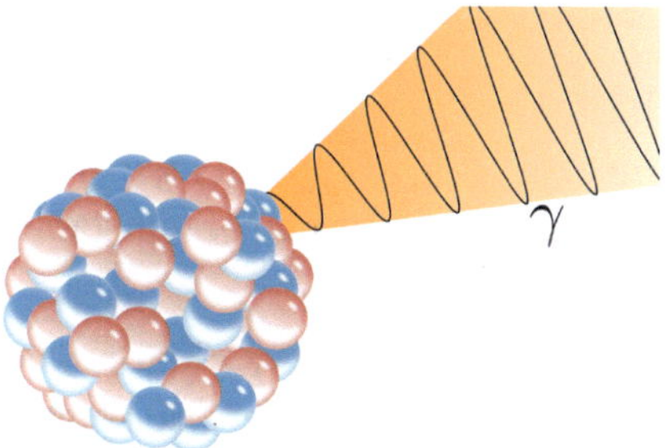

Figure 6.6. Schematic of the Gamma Rays (Source: Inductive load, Creative Commons License).

The nomenclature "gamma ray" was introduced by the esteemed British physicist Ernest Rutherford in the year 1903 subsequent to his initial investigations into the emanations of radioactive nuclei. Similar to the discrete energy levels exhibited by atoms due to the various arrangements of orbiting electrons, atomic nuclei possess energy level structures that are set by the configurations of the constituent protons and neutrons. The energy disparities observed among atomic energy levels generally range from 1 to 10 electron volts (eV), whereas energy disparities within nuclei typically span from 1 kilo-electron volt (keV) to 10 mega-electron volts (MeV). When a nucleus undergoes a transition from a state of higher energy to a state of lower energy, it releases a photon to dissipate the surplus energy. The disparities in nuclear energy levels are associated with wavelengths of photons falling within the gamma-ray region (Magdziarz & Zdziarski, 1995).

When an atomic nucleus that is in an unstable state undergoes decay, resulting in the formation of a stabler nucleus (as observed in radioactivity), it is possible for the resultant "daughter" nucleus to be generated in an energetically excited state. The subsequent transition of the daughter nucleus to a state of lower energy leads to the release of a gamma-ray photon. The technique of gamma-ray spectroscopy entails the accurate determination of the energy levels of gamma-ray photons emitted by various atomic nuclei. This method enables the characterization of nuclear energy-level configurations and facilitates the detection and identification of minute quantities of radioactive elements based on their gamma-ray emissions. Gamma rays are generated through the significant phenomenon known as pair annihilation, wherein an electron and its corresponding antiparticle, a positron, annihilate each other, resulting in the creation of two photons. The emission of photons occurs in diametrically opposed directions, with each photon possessing an energy of 511 keV. This energy corresponds to the rest mass energy, as per the concept of relativistic mass, of both the electron as well as positron. Gamma rays are additionally generated through the process of decay in certain subatomic particles that are unstable, like the neutral pion (Fagg & Hanna, 1959).

Gamma-ray photons, similar to X-ray photons, belong to the category of ionizing radiation. When they traverse a substance, their typical mode of energy transfer involves the liberation of electrons from molecules and atoms. In the lower energy spectra, it is common for an atom to fully absorb a gamma-ray photon, resulting in the transfer of the gamma ray's energy to a solitary ejected electron, as observed in the photoelectric effect. Gamma rays with higher energy exhibit a greater propensity to undergo scattering interactions with atomic electrons, resulting in the transfer of a portion of their energy during each scattering event (referred to as the Compton Effect). The detection of gamma rays commonly relies on the observation of the interactions between gamma rays and atomic electrons in various mediums such as gases, crystals, and semiconductors. This approach is well-established and extensively utilized in radiation measurement as well as scintillation counter techniques.

In addition to their interaction with other forms of matter, gamma rays are capable of interacting with atomic nuclei. During the pair production process, a gamma-ray photon with an energy that surpasses two times the rest mass energy of the electron (which is greater than 1.02 MeV) undergoes direct conversion into an electron-positron pair when it comes into close proximity with a nucleus, as depicted in the accompanying photograph. At energy levels exceeding 10 MeV, it is possible for a gamma ray to undergo direct absorption by a nucleus, resulting in the expulsion of nuclear particles (referred to as photodisintegration) or the division of the nucleus, a phenomenon known as photofission.

Gamma rays have found significant utility in the field of medicine, particularly in two key areas: positron emission tomography (PET) imaging and radiation therapy for the treatment of cancerous tumors. During a PET scan, a transient radioactive pharmaceutical that emits positrons is administered intravenously. The selection of this pharmaceutical is based on its involvement in a specific physiological process, such as brain function. Positrons that are emitted undergo rapid combination with adjacent electrons, resulting in the production of two gamma rays with energies of 511 keV each, which propagate in opposite directions as a consequence of pair annihilation. Following the identification of gamma rays, a generated by computers reconstruction is employed to visualize the spatial distribution of gamma-ray emissions, thereby generating an image that effectively delineates the exact position of the biological process under investigation (Piran, 2005).

Gamma rays, being a form of ionizing radiation that possesses high penetrating capabilities, induce notable biochemical alterations in living cells, as evidenced by the occurrence of radiation injury. Radiation therapies exploit this characteristic to specifically eradicate malignant cells within confined, localized tumors. Radioactive isotopes are introduced into the vicinity of the tumor either through injection or implantation. Subsequently, the emitted gamma rays from the radioactive nuclei perpetually bombard the affected region, thereby impeding the proliferation of the malignant cells.

Airborne surveys aimed at detecting gamma-ray emissions originating from the Earth's surface are conducted with the purpose of identifying mineral deposits that harbor minute quantities of radioactive elements, notably uranium and thorium. The utilization of both aerial and ground-based gamma-ray spectroscopy is employed as a means of providing support for geologic mapping, mineral exploration, and the identification of environmental contamination. The detection of gamma rays originating from celestial bodies was initially accomplished during the 1960s, subsequently leading to the establishment of gamma-ray astronomy as a firmly established area of scientific inquiry. Similar to the examination of astronomical X-rays, the observation of gamma-rays necessitates conducting observations beyond Earth's atmosphere, which is highly absorptive. This is typically achieved through the utilization of orbiting satellites and high-altitude balloons (refer to telescope: Gamma-ray telescopes). Numerous enigmatic and inadequately comprehended celestial gamma-ray sources exist, encompassing potent focal sources provisionally recognized as pulsars, quasars, as well as supernova remnants. One of the most intriguing unexplained astronomical phenomena pertains to gamma-ray bursts, which are characterized by brief and highly intense emissions originating from sources that appear to be distributed uniformly across the sky (Hooper & Linden, 2011).

6.9. GRAVITATIONAL WAVES

Gravitational waves manifest as perturbations in the fabric of space-time, resulting from exceedingly forceful and energetic phenomena occurring within the cosmos. In 1916, Albert Einstein made a prediction regarding the presence of gravitational waves within the framework of the general idea of relativity. According to Einstein's mathematical framework, it has been demonstrated that objects with significant mass and acceleration, such as neutron stars and black holes in mutual orbit, can induce disturbances in the fabric of space-time. These disturbances manifest as propagating waves of oscillating space-time, emanating outward from the source. The propagation of these cosmic ripples would take place at the velocity of light, conveying valuable data regarding their sources, as well as providing insights into the fundamental characteristics of gravity (Fryer & New, 2011).

Did You Know?

Gravitational waves are ripples in the fabric of spacetime caused by massive celestial events, such as the collision of black holes or neutron stars. Their detection in 2015 confirmed a major prediction of Albert Einstein's theory of general relativity, opening up a new way to observe and study the universe.

Catastrophic phenomena, such as the collision of black holes, the explosion of massive stars known as supernovae during their final stages, and the collision of neutron stars, generate gravitational waves of the highest magnitude (Figure 6.7). Additional gravitational waves are hypothesized to originate from the rotational motion of neutron stars exhibiting non-spherical shapes, and conceivably, from the remnants of gravitational radiation generated during the primordial event known as the Big Bang (Sazhin, 1978).

Figure 6.7. Schematic of the gravitational waves (Source: Symmetry Magazine, Creative Commons License).

The provided animation effectively demonstrates the emission of gravitational waves by a binary system of neutron stars during their orbital motion and subsequent merger. The credit for this visualization goes to NASA's Goddard Space Flight Center. It should be noted that gravitational waves possess an inherent quality of being imperceptible to the human eye. The purpose of their visibility in this context is to demonstrate their dissemination from the origin.

Despite Albert Einstein's prediction of the presence of gravitational waves in 1916, empirical evidence confirming their presence was not obtained until

1974, two decades subsequent to his demise. During that particular year, a binary pulsar located approximately 21,000 light years away from Earth was discovered by astronomers Russell Hulse as well as Joseph Taylor, who conducted their observations at the Arecibo Radio Observatory situated in Puerto Rico. The system in question aligns precisely with the theoretical framework of general relativity, which posits the emission of gravitational waves. With the intention of examining Einstein's prediction, Taylor, along with his colleagues Joel Weisberg and Lee Fowler, initiated the monitoring of radio emissions emitted by stars in order to quantify the variations in their orbital period (Cervantes-Cota et al., 2016).

After a span of merely four years, the initial observations indicated a discernible alteration in the period, providing confirmation that the stars were gradually approaching each other in accordance with the predictions of general relativity (GR), assuming the emission of gravitational waves. The observed rate of convergence closely aligned with the rate anticipated by GR, exhibiting a remarkable agreement within a margin of 0.5%. The Nobel Prize in Physics was awarded to Hulse and Taylor in 1993 for their groundbreaking discovery of a novel pulsar, which has significantly expanded the potential for gravitational research.

Subsequently, numerous astronomers have conducted investigations on pulsar radio emissions, which emanate from pulsars, known as neutron stars that emit radio wave beams. These studies have yielded comparable outcomes, thereby providing additional substantiation for the presence of gravitational waves. However, these confirmations have consistently been obtained indirectly or through mathematical calculations rather than through direct interaction or contact with an instrument (Einstein & Rosen, 1937).

A significant shift occurred on September 14, 2015, when the Laser Interferometer Gravitational-Wave Observatory (LIGO) successfully detected the oscillations in the fabric of spacetime resulting from the interaction of gravitational waves produced by the merger of two black holes situated approximately 1.3 billion light-years distant. The initial detection made by the Laser Interferometer Gravitational-Wave Observatory (LIGO) will undoubtedly be recorded as a momentous scientific milestone in the annals of human history.

The generation of gravitational waves is characterized by highly turbulent and disruptive processes. However, upon their arrival at Earth, these waves undergo a significant reduction in magnitude, diminishing by a factor of thousands of billions. This diminishment occurs as the waves propagate through both temporal and spatial dimensions, analogous to the gradual decrease in amplitude observed in waves emanating from a pebble dropped into a body of water as they propagate away from the point of origin. Indeed, upon the arrival of gravitational waves from the initial detection made by LIGO, the magnitude of the resultant perturbation in the fabric of space-time was found to be approximately one thousand times smaller than the dimensions of an atomic nucleus. The primary objective of the Laser Interferometer Gravitational-Wave Observatory (LIGO) was to facilitate the attainment of exceptionally precise measurements at an unimaginably minuscule scale (Abbott et al., 2016).

Activity 6.1

Explain the applications of microwaves in both communication and radar systems. Provide examples of how microwaves are utilized in modern technologies to enhance communication and facilitate radar-based operations.

SUMMARY

The chapter delves into the study of waves that extend beyond the conventional domains of light and sound. The subject matter encompasses a broad range of wave phenomena, encompassing electromagnetic waves, microwaves, X-rays, gamma rays, waves from water, seismic waves, and gravitational waves. These waves possess a wide range of applications, encompassing areas such as communication, medical imaging, as well as the comprehension of Earth's interior as well as celestial phenomena. The chapter elucidates the impact of these waves on our comprehension of the cosmos and demonstrates their importance in diverse domains of scientific inquiry and technological advancement. The readers are provided with a deeper understanding of the concealed realm of waves, thereby developing an enhanced appreciation for their significant influence on both the natural environment and other domains.

REVIEW QUESTIONS

1. How are electromagnetic (EM) waves formed, and what is their mathematical representation? Describe the electromagnetic wave equation.
2. What are some practical applications of electromagnetic waves in

various fields, such as communication, radar, and microwave heating?

3. Discuss the formation and characteristics of water waves. What distinguishes a water wave tsunami from regular water waves?
4. How are radio waves generated, and what are their common uses in modern technology?
5. Explain the characteristics of X-rays and their applications in different areas of science and medicine.
6. What are gamma rays, and how do they differ from X-rays? Explore their significance and applications in various fields.

REFERENCES

1. Abbott, B. P., Abbott, R., Abbott, T., Abernathy, M. R., Acernese, F., Ackley, K., ... & Cavalieri, R. (2016). Observation of gravitational waves from a binary black hole merger. *Physical Review Letters*, *116*(6), 061102.
2. Adamovich, S., Burlachkova, N. I., & Fel'dman, A. G. (1984). Wave nature of the central process of formation of the trajectories of change in the joint angle in man. *Biophysics*, *29*(1), 130–134.
3. Ambrose, B. S., Heron, P. R., Vokos, S., & McDermott, L. C. (1999). Student understanding of light as an electromagnetic wave: Relating the formalism to physical phenomena. *American Journal of Physics*, *67*(10), 891–898.
4. Annenkov, S. Y., & Shrira, V. I. (2006). Direct numerical simulation of downshift and inverse cascade for water wave turbulence. *Physical Review Letters*, *96*(20), 204501.
5. Armstrong, J. A., Bloembergen, N., Ducuing, J., & Pershan, P. S. (1962). Interactions between light waves in a nonlinear dielectric. *Physical Review*, *127*(6), 1918.
6. Baylis, C., Fellows, M., Cohen, L., & Marks II, R. J. (2014). Solving the spectrum crisis: Intelligent, reconfigurable microwave transmitter amplifiers for cognitive radar. *IEEE Microwave Magazine*, *15*(5), 94–107.
7. Bergé, L. (1998). Wave collapse in physics: principles and applications to light and plasma waves. *Physics Reports*, *303*(5–6), 259–370.
8. Brett, J. M., & Krelle, A. (2018). A study of bubble collapse pressure pulse waves from small scale underwater explosions near the water surface. *Journal of Sound and Vibration*, *435*, 91–103.
9. Brown, W. C. (1984). The history of power transmission by radio waves. *IEEE Transactions on Microwave Theory and Techniques*, *32*(9), 1230–1242.
10. Buttgen, B., Oggier, T., Kaufmann, R., Seitz, P., & Blanc, N. (2004, April). Demonstration of a novel drift field pixel structure for the demodulation of modulated light waves with application in three-dimensional image capture. In *Three-Dimensional Image Capture and Applications VI*. SPIE. (Vol. 5302, pp. 9–20).
11. Cervantes-Cota, J. L., Galindo-Uribarri, S., & Smoot, G. F. (2016). A brief history of gravitational waves. *Universe*, *2*(3), 22.

12. Chen, S., Li, Z., Zhang, Y., Cheng, H., & Tian, J. (2018). Phase manipulation of electromagnetic waves with metasurfaces and its applications in nanophotonics. *Advanced Optical Materials*, *6*(13), 1800104.
13. Constant, T., Moyne, C., & Perre, P. (1996). Drying with internal heat generation: theoretical aspects and application to microwave heating. *AIChE Journal*, *42*(2), 359–368.
14. Cox, S., Lashley, J. C., Rosten, E., Singleton, J., Williams, A. J., & Littlewood, P. B. (2007). Evidence for the charge-density-wave nature of the stripe phase in manganites. *Journal of Physics: Condensed Matter*, *19*(19), 192201.
15. Dacey, G. C. (1962). Optical Masers in Science and Technology: Advances in the control of light waves give promise of important applications in science and technology. *Science*, *135*(3498), 71–74.
16. Donnet, B. J., & Longstaff, I. D. (2006, September). Combining MIMO radar with OFDM communications. In *2006 European Radar Conference*. IEEE. (Vol. 2, pp. 37–40).
17. Einstein, A., & Rosen, N. (1937). On gravitational waves. *Journal of the Franklin Institute*, *223*(1), 43–54.
18. Elkhoury, J. E., Brodsky, E. E., & Agnew, D. C. (2006). Seismic waves increase permeability. *Nature*, *441*(7097), 1135–1138.
19. Fagg, L. W., & Hanna, S. S. (1959). Polarization measurements on nuclear gamma rays. *Reviews of Modern Physics*, *31*(3), 711.
20. Fryer, C. L., & New, K. C. (2011). Gravitational waves from gravitational collapse. *Living Reviews in Relativity*, *14*, 1–93.
21. Hall, H. (1936). The theory of photoelectric absorption for x-rays and γ-rays. *Reviews of Modern Physics*, *8*(4), 358.
22. Havelock, T. H. (1908). The propagation of groups of waves in dispersive media, with application to waves on water produced by a travelling disturbance. *Proceedings of the Royal Society of London. Series A, Containing Papers of a Mathematical and Physical Character*, *81*(549), 398–430.
23. Hennino, R., Trégourès, N., Shapiro, N. M., Margerin, L., Campillo, M., Van Tiggelen, B. A., & Weaver, R. L. (2001). Observation of equipartition of seismic waves. *Physical Review Letters*, *86*(15), 3447.
24. Henriksen, E. K., Angell, C., Vistnes, A. I., & Bungum, B. (2018). What is light? Students' reflections on the wave-particle duality of light and

the nature of physics. *Science & education*, *27*, 81–111.

25. Hikita, M., Ohtsuka, S., Teshima, T., Okabe, S., & Kaneko, S. (2007). Electromagnetic (EM) wave characteristics in GIS and measuring the EM wave leakage at the spacer aperture for partial discharge diagnosis. *IEEE Transactions on Dielectrics and Electrical Insulation*, *14*(2), 453–460.
26. Honda, H. (1962). Earthquake mechanism and seismic waves. *Journal of Physics of the Earth*, *10*(2), 1–97.
27. Hooper, D., & Linden, T. (2011). Origin of the gamma rays from the Galactic Center. *Physical Review D*, *84*(12), 123005.
28. Hui, S., Zhang, L., & Wu, H. (2023). Cationic doping induced sulfur vacancy formation in polyionic sulfide for enhanced electromagnetic wave absorption. *Journal of Colloid and Interface Science*, *629*, 147–155.
29. Jansky, K. G. (1933). Radio waves from outside the solar system. *Nature*, *132*(3323), 66–66.
30. Jian, S., & Changlong, G. (2006). Parameterized first-guess spectrum method for retrieving directional spectrum of swell-dominated waves and huge waves from SAR images. *Chinese Journal of Oceanology and Limnology*, *24*(1), 12–20.
31. Justesen, D. R. (1975). Microwaves and behavior. *American Psychologist*, *30*(3), 391.
32. Kane, P. P., Kissel, L., Pratt, R. H., & Roy, S. C. (1986). Elastic scattering of γ-rays and X-rays by atoms. *Physics Reports*, *140*(2), 75–159.
33. Kiselev, A. P. (2007). Localized light waves: Paraxial and exact solutions of the wave equation (a review). *Optics and Spectroscopy*, *102*, 603–622.
34. Korovin, S. D., Rostov, V. V., Polevin, S. D., Pegel, I. V., Schamiloglu, E., Fuks, M. I., & Barker, R. J. (2004). Pulsed power-driven high-power microwave sources. *Proceedings of the IEEE*, *92*(7), 1082–1095.
35. Lagos, L. E., Li, W., Ebadian, M. A., White, T. L., Grubb, R. G., & Foster, D. (1995). Heat transfer within a concrete slab with a finite microwave heating source. *International Journal of Heat and Mass Transfer*, *38*(5), 887–897.
36. Li, C., Peng, Z., Huang, T. Y., Fan, T., Wang, F. K., Horng, T. S., ... & Lin, J. (2017). A review on recent progress of portable short-

range noncontact microwave radar systems. *IEEE Transactions on Microwave Theory and Techniques*, *65*(5), 1692–1706.

37. Longuet-Higgins, M. S. (1953). Mass transport in water waves. *Philosophical Transactions of the Royal Society of London. Series A, Mathematical and Physical Sciences*, *245*(903), 535–581.
38. Magdziarz, P., & Zdziarski, A. A. (1995). Angle-dependent Compton reflection of X-rays and gamma-rays. *Monthly Notices of the Royal Astronomical Society*, *273*(3), 837–848.
39. Marshall, T., & Santos, E. (1988). Stochastic optics: A reaffirmation of the wave nature of light. *Foundations of physics*, *18*, 185–223.
40. Masood, W., Eliasson, B., & Shukla, P. K. (2010). Electromagnetic wave equations for relativistically degenerate quantum magnetoplasmas. *Physical Review E*, *81*(6), 066401.
41. Meinel, H. H. (1995). Commercial applications of millimeterwaves: history, present status, and future trends. *IEEE Transactions on Microwave Theory and Techniques*, *43*(7), 1639–1653.
42. Piran, T. (2005). The physics of gamma-ray bursts. *Reviews of Modern Physics*, *76*(4), 1143.
43. Ramadan, K. T., Hassan, H. S., & Hanna, S. N. (2011). Modeling of tsunami generation and propagation by a spreading curvilinear seismic faulting in linearized shallow-water wave theory. *Applied Mathematical Modelling*, *35*(1), 61–79.
44. Ramaty, R., Kozlovsky, B., & Lingenfelter, R. E. (1975). Solar gamma rays. *Space Science Reviews*, *18*(3), 341–388.
45. Rosen, A., Stuchly, M. A., & Vander Vorst, A. (2002). Applications of RF/microwaves in medicine. *IEEE Transactions on Microwave Theory and Techniques*, *50*(3), 963–974.
46. Sahoo, M., & Chakraverty, S. (2022). Sawi transform based homotopy perturbation method for solving shallow water wave equations in fuzzy environment. *Mathematics*, *10*(16), 2900.
47. Santacruz, J. P., Rommel, S., Johannsen, U., Jurado-Navas, A., & Monroy, I. T. (2020). Analysis and compensation of phase noise in mm-wave OFDM ARoF systems for beyond 5G. *Journal of Lightwave Technology*, *39*(6), 1602–1610.
48. Sazhin, M. V. (1978). Opportunities for detecting ultralong gravitational waves. *Sov. Astron*, *22*, 36–38.

49. Sobol, H., & Tomiyasu, K. (2002). Milestones of microwaves. *IEEE Transactions on Microwave Theory and Techniques*, *50*(3), 594–611.
50. Stockman, H. (1948). Communication by means of reflected power. *Proceedings of the IRE*, *36*(10), 1196–1204.
51. Sun, Q., Ji, Y., He, L., & Long, X. (2020). In situ formation of Fe 3 O 4/N-doped carbon coating on the surface of carbon fiber with improved electromagnetic wave-absorption property. *RSC Advances*, *10*(51), 30443–30450.
52. Taniguchi, S., Arima, T., Ruggeri, T., & Sugiyama, M. (2014). Thermodynamic theory of the shock wave structure in a rarefied polyatomic gas: Beyond the Bethe-Teller theory. *Physical Review E*, *89*(1), 013025.
53. Thurber, K. R., Yau, W. M., & Tycko, R. (2010). Low-temperature dynamic nuclear polarization at 9.4 T with a 30 mW microwave source. *Journal of Magnetic Resonance*, *204*(2), 303–313.
54. Van de Hulst, H. C. (1945). Radiogolven uit het wereldruim: II. Herkomst der radiogolven; Radiogolven uit het wereldruim: II. Herkomst der radiogolven; Radio waves from space. *Nederlandsch Tijdschrift voor Natuurkunde*, *11*, 210–221.
55. Wu, W., Liu, Y. L., Zhang, A. M., Liu, N., & Liu, L. (2020). Numerical investigation on underwater explosion cavitation characteristics near water wave. *Ocean Engineering*, *205*, 107321.
56. Yuan, K., Han, D., Liang, J., Zhao, W., Li, M., Zhao, B., ... & Zhang, R. (2021). Microwave induced in-situ formation of SiC nanowires on SiCNO ceramic aerogels with excellent electromagnetic wave absorption performance. *Journal of Advanced Ceramics*, *10*, 1140–1151.
57. Zhang, J., & Northoff, G. (2022). Beyond noise to function: reframing the global brain activity and its dynamic topography. *Communications Biology*, *5*(1), 1350.

CHAPTER 7

Wave Technology and Its Applications

Contents

7.1. INTRODUCTION

Waves have found a wide range of applications in numerous disciplines, revolutionizing the way we communicate and engage with technology. Communication technology extensively depends on the utilization of waves, particularly electromagnetic waves, which facilitate wireless communication, satellite communication, as well as internet connectivity via radio waves, microwaves, as well as fiber optics. Furthermore, the field of medical imaging and diagnosis has been significantly transformed by the utilization of waves. Ultrasound technology, for instance, has enabled the acquisition of real-time images of internal structures, while X-rays have facilitated the visualization of bones as well as dense tissues (Ratnam et al., 2020). The utilization of concentrated gamma ray beams for the purpose of cancer treatment is known as gamma ray therapy. Sonar technology involves the utilization of sound waves to facilitate the navigation and detection of submerged objects. Conversely, remote sensing employs infrared as well as microwave waves to support various applications such as forecasting the weather, agriculture, as well as disaster management. Acoustic levitation as well as optical communication utilizing light waves possess a multitude of applications within the fields of medicine and information transmission. Radio waves are utilized in the field of space exploration and astronomy to facilitate the examination of celestial entities, whereas gamma rays are employed to detect and analyze high-energy phenomena. Waves play a crucial role in various fields such as materials science, producing goods, seismic monitoring, as well as entertainment, exerting a significant influence on our society and fostering advancements in multiple domains (Choi et al., 2009).

7.2. ELECTROMAGNETIC WAVES AND COMMUNICATION TECHNOLOGY

The act of greeting someone is commonly expressed through the physical gesture of waving one's hands. However, it is important to note that there exist numerous alternative methods of communication. Sound waves are a means by which communication between individuals is facilitated. During the process of speech production, the passage of air through the vocal cords generates vibrations that propagate through the surrounding air medium, ultimately reaching the auditory system within the human ear. Nevertheless, the propagation of sound waves is limited by their gradual attenuation over distance.

Historically, long-distance communication was characterized by sluggishness and inefficiency due to the manual delivery of messages. However, as a result of technological progress, novel methods of communication have been discovered, employing electromagnetic waves. These waves are a constituent of the electromagnetic spectrum as well as facilitated long-range communication. Certain techniques can also be advantageous when applied to limited distances. The utilization of these waves has facilitated the enhancement of communication in terms of speed and reliability (Cella et al., 2009).

Please refer to the diagram provided below, illustrating the electromagnetic spectrum. This phenomenon serves as a means of illustrating the dispersion of various forms of electromagnetic radiation in relation to their respective frequencies and wavelengths. Waves characterized by higher frequencies, like gamma rays and x-rays, exhibit shorter wavelengths and possess greater energy. Conversely, microwaves as well as radio waves exhibit the least frequencies and energies, while concurrently possessing the longest wavelengths.

The comprehension of the electromagnetic spectrum holds significance as it facilitates the comprehension of the utilization of various wave types in our everyday existence. An illustration of this can be seen in the utilization of radio waves for communication purposes, as opposed to the application of x-rays in the field of medical imaging (Figure 7.1). It is remarkable to consider the pervasive existence of these waves in our surroundings, despite their imperceptibility to our senses.

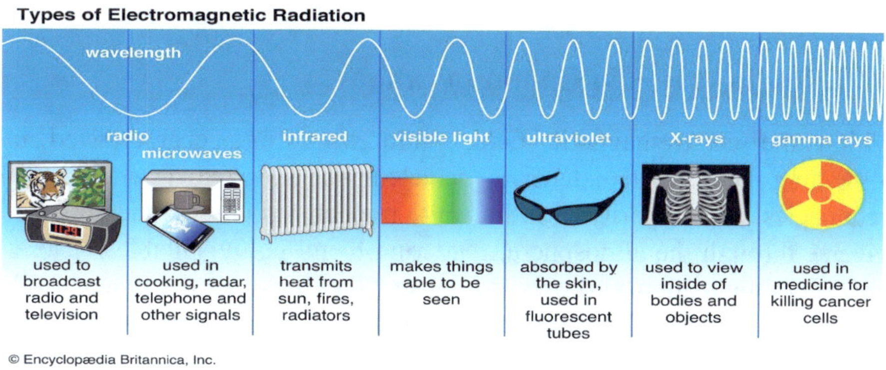

Figure 7.1. Types of electromagnetic radiation (Source: Britannica, Creative Commons License).

The various segments of the electromagnetic spectrum are arranged in a systematic manner based on their respective frequencies and wavelengths (Cheng et al., 2022).

Electromagnetic waves possess significant utility in the transmission and reception of messages. The various segments of the electromagnetic spectrum exhibit distinct energy levels, frequencies, and wavelengths, thereby rendering certain segments more suitable for specific applications than others. In the realm of telecommunications, it is customary to utilize solely the lower frequency components of the spectrum. This is due to the limited efficacy of high-frequency radiation, such as ultraviolet light, in long-distance applications. Researchers are currently investigating the potential application of ultraviolet light in the field of communication. Despite the absorption of UV light by the Earth's atmosphere, it is capable of propagating over considerable distances, spanning several miles, given its origin from a sufficiently intense source. Utilizing ultraviolet light as a means of communication offers two primary advantages. Firstly, due to its elevated frequency, it possesses the capacity to transmit a greater amount of information per unit of time compared to waves of lower frequencies such as visible light or infrared. Furthermore, it is important to note that airborne particles possess the capability to both absorb and disperse ultraviolet (UV) light, thereby facilitating the dispersion of a signal across a significant spatial extent. Moreover, due to the phenomenon of UV light's ability to undergo multiple reflections, it has the potential to serve as a means of communication between two locations that lack direct line of sight (Wang et al., 2020).

7.2.1. Radio Waves in Communications

Within the electromagnetic field, radio waves are characterized by possessing the most extensive wavelengths and the least elevated frequencies. The majority of radio wave frequencies possess the ability to traverse through the atmosphere, rendering them highly suitable for long-range communication purposes. Moreover, they traverse the human body without causing any adverse effects. The radio spectrum encompasses a vast expanse. The electromagnetic spectrum encompasses wavelengths ranging from 1 millimeter to 10,000 kilometers, surpassing the Earth's diameter in magnitude. Radio waves of varying wavelengths and velocities exhibit distinct behaviors. Typically, these waves are employed for the purpose of transmitting messages, such as in the context of television and radio broadcasts. Oscillatory electrical circuits facilitate the transmission and reception of radio waves. When a radio wave penetrates a conductor, it induces an alternating current. The velocity of the radio wave is equivalent to that of the alternating current. This paper elucidates the utilization of computers for encoding information prior to its transmission via radio waves, as well as the subsequent decoding process facilitated by computers upon the wave's reception (Raza & Salam, 2020).

Short-wavelength waves are referred to as radio waves with an overall length of below approximately 3 MHz. These radio waves are employed for the transmission of data via two antennas that are within visual range of one another. This implies that the propagation of the wave is limited to the visual range, approximately 64 km, within the Earth's surface. Nevertheless, the magnitude of the waves is sufficient to penetrate the majority of structures and vegetation without encountering any significant obstruction. High frequencies are typically employed in situations where a substantial amount of information needs to be transmitted simultaneously, such as in the utilization of radio, television, cellular devices, or radar systems. Radio waves with frequencies below 3 MHz exhibit a phenomenon wherein they undergo bending as they propagate beyond the horizon. This enables radios to transmit signals across vast distances, adhering to the curvature of the Earth. The propagation distance of the signal will increase when the frequency is decreased. Regrettably, the transmission capacity decreases as the frequency decreases. Radio waves within the frequency range of 3 MHz to 30 MHz are susceptible to scattering by ionized particles present in the ionosphere, subsequently resulting in their redirection towards the Earth's surface. The phenomenon referred to as "skywave propagation" enables the transmission of radio waves across different continents. Presently, the utilization of this particular mode of communication has become infrequent due to the inherent unpredictability of weather conditions (Moridi et al., 2014).

7.2.2. Visible Light Waves in Communications

Indeed, it is worth noting that visible light serves as a means of natural communication among individuals (Figure 7.2). Consequently, our focus will be solely on the utilization of visible light as a technology for communication purposes. The semaphore telegraph system was invented in 1792 by Claude Chappe, a French individual. The semaphore system was an early and rudimentary method employed for the transmission of information over long distances through the use of visual signals. Tall structures were constructed in such a way that they maintained direct visual contact with one another, with the specific distances ranging from 5 to 20 miles, contingent upon the surrounding topography. Each tower would be manned by a human operator equipped with a spyglass, diligently observing the adjacent towers for any signals. Upon receipt of a message, the operator would subsequently transmit the message to the subsequent tower. The aforementioned technology served as a starting point to the electrical telegraph, which eventually superseded the semaphore system within a span of approximately 100 years.

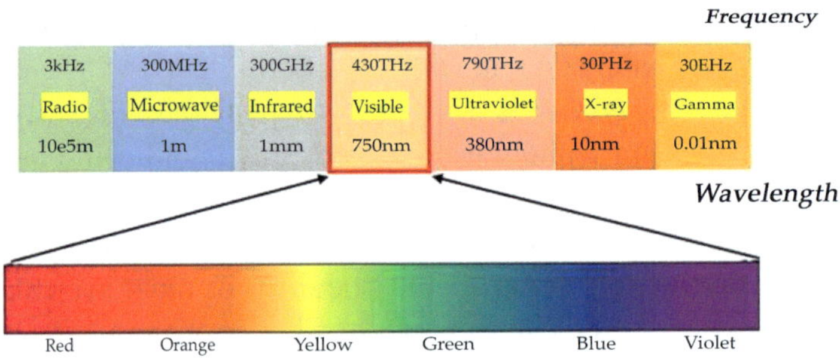

Figure 7.2. Visible Light Communication Frequency Spectrum (Source: Akashkumar Rajaram, Creative Commons License).

In contemporary times, the transmission of signals through the utilization of fiber optic cables is achieved by employing visible light. Fiber optic cables have the capability to transmit high-speed internet connectivity. Fiber optic cables consist of flexible glass fibers that enable the internal reflection of electromagnetic waves, facilitating their transmission via the cable at the velocity of light. The cables are securely encased within a plastic sheath in order to provide protection for the fragile glass fibers contained within. Optical fiber cables possess the capability to transmit signals over extensive distances, primarily owing to the minimal signal loss incurred through scattering and absorption of the wave during reflection at the cable boundaries. The utilization of higher quality glass in the cable facilitates reduced attenuation as the wave propagates (Wu et al., 2014).

7.2.3. Infrared Waves in Communications

The range of electromagnetic radiation known as infrared light spans from 700 nm (0.0007 mm) to 1 mm. In the realm of communications, infrared transmissions find predominant utilization within the context of short-range applications. The TV remote control serves as a prevalent illustration of this phenomenon. One primary limitation associated with infrared communications pertains to the inability of infrared waves to penetrate solid objects, resulting in a restricted transmission range. Additionally, the atmosphere partially absorbs these waves, further contributing to their limited reach. In certain circumstances, such as within the confines of one's personal residence, this can be transformed into a favorable aspect. The presence of an infrared television remote in the living room does not pose interference to analogous devices located in other rooms within the household due to the inability of infrared signals to penetrate walls. In addition, it is noteworthy that infrared transmitters exhibit a comparatively affordable manufacturing process and possess minimal power demands, thereby ensuring cost-effectiveness for end-users.

In the context of fiber optic cables, infrared wavelengths are employed to mitigate signal loss. The scattering or absorption of visible light by impurities within fiber optic cables is more pronounced compared to infrared light, resulting in increased signal loss during transmission through the cable. Nonetheless, the data transfer rate in the infrared spectrum is comparatively lower than that in the visible spectrum, primarily attributed to the reduced frequencies of infrared light (Zhang et al., 2019).

7.2.4. Satellite Communications

One method of achieving global communication involves the utilization of satellites, which are frequently used by satellite television companies as well as mobile phone networks (Figure 7.3). For example, a cellular base station facilitates the transmission and reception of data between numerous mobile devices in close proximity through the utilization of radio frequency electromagnetic waves. The mobile phone tower will subsequently establish communication with a satellite positioned in geostationary orbit through the utilization of microwaves. A geostationary satellite is designed to maintain a fixed position relative to a specific point on the Earth's surface, enabling continuous and uninterrupted interaction with ground-based equipment throughout the day and night. The utilization of the microwave spectrum is advantageous due to its ability to penetrate the Earth's atmosphere without obstruction, enabling the transmission of a greater volume of data per unit of time compared to radio waves, owing to their elevated frequencies.

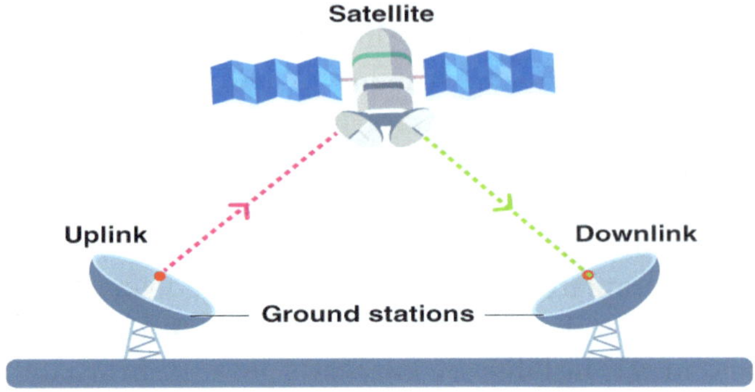

Figure 7.3. Schematic of the satellite communication (Source: BYJU's, Creative Commons License).

The satellite has the capability to establish communication with other mobile phone towers located within a substantial geographical region on the Earth's surface (Figure 7.4). In the realm of commercial satellite communication, direct inter-satellite communication is a seldom occurrence, as signals necessitate relaying to terrestrial receivers (Kodheli et al., 2020).

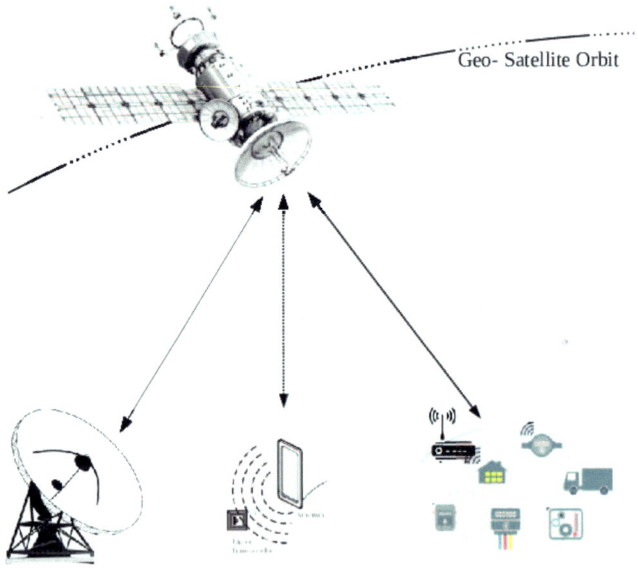

Figure 7.4. Using a satellite to communicate over long distances (Source: Mohammed Riyaz Ahmed, Creative Commons License).

The establishment of a geostationary orbit may be accomplished by deploying a satellite into a trajectory positioned above the Earth's equator, at an approximate altitude of 36,000 kilometers from the planet's surface. At this altitude, a satellite will require precisely 24 hours (23 hours and 56 minutes) to go through a single revolution. If the satellite's orbit aligns with the Earth's rotational direction, it will exhibit a stationary position in the Earth's sky as observed from the surface.

Key Takeaways: Waves in Communication the study of waves in communication is of significant importance in the field of telecommunications. Waves play a crucial role in transmitting information across various mediums, such as radio, television, and the internet. Understanding the properties and characteristics of waves is essential for designing efficient communication systems. One key takeaway is that A wide range of light wavelengths within the electromagnetic spectrum are utilized for the purpose of communication. The transmission capacity of a wave increases proportionally with its frequency, allowing for a greater amount of data or information to be conveyed within a given time frame. The practicality of utilizing electromagnetic radiation using frequencies higher than those of visible light diminishes progressively (Maral et al., 2020). Radio waves possess significantly long wavelengths and are commonly employed in the transmission of television and radio signals. At the lowest frequencies, radio waves have the capability to be transmitted over distances of several hundred miles. Conductors possess the ability to assimilate and propagate radio waves, thereby facilitating the electronic encoding and decoding of information. Fiber optic cables facilitate the transmission of data through the utilization of either visible or infrared light. These devices possess utility due to their high data transmission rate and minimal signal loss. The infrared spectrum is primarily advantageous for applications involving short-range communications, like those found in remote control devices.

The atmosphere absorbs a portion of infrared radiation, while obstacles completely obstruct its transmission. Satellites facilitate global-scale communications. Microwave radiation is employed owing to its comparatively higher data transfer rate in comparison to radio waves. Furthermore, it should be noted that microwaves possess the characteristic of being visible to the Earth's the environment. The majority of communication satellites are positioned in geostationary orbits in order to ensure uninterrupted transmission between the satellites and ground-based equipment (De Sanctis et al., 2015).

7.3. MEDICAL APPLICATIONS OF WAVES

The utilization of waves in the medical field has brought about a significant transformation, offering highly effective resources for the purposes of imaging, diagnosis, as well as treatment. This section delves into three notable uses of waves: Ultrasound Technology as well as Imaging, X-rays as well as Radiography, and Gamma Ray Treatment in Medicine (Topfer & Oberhammer, 2015).

7.3.1. Ultrasound Technology and Imaging

The utilization of high-frequency sound waves in ultrasound technology enables the creation of intricate visual representations of internal anatomical structures. The non-invasive nature and capacity to visualize soft tissues have rendered it a fundamental component of medical imaging. Ultrasound imaging involves the emission of sound waves by a transducer into the human body. These sound waves subsequently reflect back as echoes upon encountering tissues of diverse densities. The transducer detects these echoes and subsequently processes them into real-time images that are displayed on a monitor.

Tip

Ultrasound technology is widely used in medical imaging to create real-time images of internal body structures. Understanding the principles of ultrasound and its safe application is essential in providing accurate diagnoses and guiding medical procedures.

Ultrasound imaging is extensively employed across diverse medical disciplines. The utilization of obstetric ultrasound is a prevalent practice in pregnancy for the purpose of monitoring fetal growth and identifying potential anomalies. Echocardiography is a diagnostic imaging technique used in the field of cardiology to facilitate the visualization and assessment of the anatomical structures and physiological function of the heart. Moreover, the utilization of ultrasound is instrumental in the diagnostic assessment of various medical conditions affecting the abdominal, pelvic, and musculoskeletal regions.

The inherent non-ionizing properties of ultrasound render it a secure modality for application in both pregnancy and pediatric populations, thereby establishing its indispensable role within contemporary medical practice (Moran & Thomson, 2020).

7.3.2. X-rays and Radiography

X-rays are a type of electromagnetic radiation characterized by higher energy levels compared to visible light. X-rays are employed in medical contexts for the purpose of radiography, a diagnostic technique facilitating the visualization of bones as well as dense tissues. When X-rays traverse the human body, they encounter differential absorption by distinct tissues. Dense anatomical structures, including bones, exhibit a higher capacity for X-ray absorption, resulting in their depiction as white on X-ray images. Conversely, tissues with lower density manifest as darker regions on the X-ray image.

Radiography assumes a pivotal role in the diagnosis of fractures, bone abnormalities, as well as joint pathologies. In the field of dentistry, this tool holds significant value as it aids in the detection of dental abnormalities. Similarly, in the context of chest X-rays, it serves as a valuable instrument for the examination of pulmonary conditions.

Nevertheless, it is imperative to exercise caution when employing X-rays owing to their ionizing properties, which have the potential to induce harm to cells and elevate the likelihood of developing cancer with prolonged exposure. In order to mitigate the potential hazards associated with X-ray imaging, it is imperative to employ appropriate shielding techniques and optimize the dosage administered (Underwood & Barbee, 1981).

7.3.3. Gamma Ray Therapy in Medicine

Gamma rays, which are the highest energy electromagnetic waves, are utilized in the medical field for the purpose of cancer treatment. Gamma ray therapy, also referred to as radiation therapy and radiotherapy, encompasses the utilization of concentrated beams of gamma rays to specifically target and eradicate malignant cells. The utilization of gamma rays, owing to their elevated energy levels, facilitates the accurate localization of malignant tumors while minimizing potential damage to adjacent healthy tissues.

The administration of gamma ray therapy may be conducted either through external or internal means. External beam radiation therapy is a treatment modality that entails the administration of gamma rays externally to the body through the utilization of a linear accelerator. The medical procedure known as internal radiation therapy, also referred to as brachytherapy, entails the insertion of a radioactive source in close proximity to or directly within the tumor (Waksman et al., 2000).

Gamma ray therapy is a crucial element in the comprehensive management of cancer, frequently employed either as a standalone intervention or in combination with surgical procedures and chemotherapeutic agents. The primary objective is to reduce the size of tumors, mitigate symptoms, and potentially achieve a cure for cancer in certain instances.

Nevertheless, it is imperative to meticulously strategize and administer gamma ray therapy in order to mitigate any potential harm to unaffected tissues and organs. Achieving optimal treatment outcomes and prioritizing patient safety necessitates the adoption of a multidisciplinary approach that encompasses the expertise of radiation oncologists, healthcare physicists, and dosimetrists.

The utilization of waves in medical contexts has revolutionized the field of healthcare, facilitating the use of non-invasive imaging techniques, precise diagnostic methods, and efficacious cancer therapies. Ultrasound technology offers significant real-time imaging of internal structures, whereas X-rays facilitate the visualization of bones and dense tissues. Gamma ray therapy is an essential component of cancer treatment as it effectively targets and eradicates malignant cells. With the ongoing advancement of technology, it is inevitable that the medical field will experience additional innovations, leading to enhanced patient care and improved outcomes by utilizing waves in contemporary medicine (Khodamoradi et al., 2022).

7.4. APPLICATIONS OF SOUND WAVES

Sound waves can be produced in various mediums, including gases such as air, as well as liquids and solids. Sound waves are classified as longitudinal waves, indicating that the oscillation of air particles occurs parallel to the direction of wave propagation (Yoneyama et al., 1983).

7.4.1. The Ear

Sound waves play a crucial role in facilitating everyday spoken communication, as they serve as the medium through which messages are transmitted to individuals (Figure 7.5). During oral communication, the articulatory movements of various components of the oral cavity generate oscillations in the adjacent air medium. The transmission of these vibrations occurs through the medium of air by perturbing adjacent air molecules. The sound wave propagates until it gets to its intended destination, typically the auditory organ of the individual with whom we are communicating.

Sound waves are comprised of compressions, wherein air molecules are compressed in closer proximity, and rarefactions, wherein the air molecules are dispersed at greater distances from each other. The following diagram depicts the regions of compression and rarefaction within a standard sound wave.

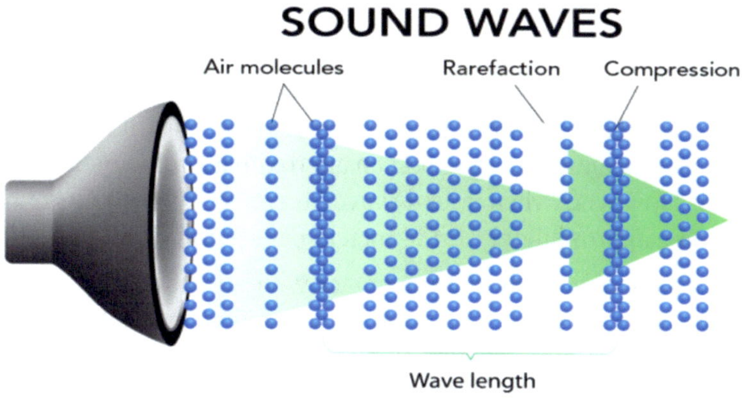

Figure 7.5. The propagation of a sound wave from its origin to the auditory organ. Regions with a higher degree of darkness exhibit a greater abundance of air molecules that are engaged in vigorous oscillatory motion, commonly referred to as compressions (Source: Harleen Saini, Creative Commons License).

The human hearing range encompasses frequency limits. Humans are unable to perceive sounds that have frequencies below or above a certain threshold. The range of frequencies in question is commonly referred to as the audible spectrum.

The velocity of a wave, denoted as v, can be determined by the product of its frequency, represented by f, and its wavelength, symbolized as λ, as expressed by the equation $v = f\lambda$ (Jeutter & Josse, 1993).

7.4.2. Sonar

Sound waves possess a multitude of applications within the domains of communication, music, as well as other disciplines pertaining to sound. Sound waves have the ability to propagate through both liquids and solids, exhibiting different properties compared to their propagation through air. Notably, water is recognized as a superior medium for sound conduction when compared to air. The aforementioned fact led to the development of the concept known as sonar, an acronym for 'sound navigation and ranging.'

Sonar technology employs acoustic waves in a liquid medium to detect and locate submerged impediments, such as naval mines or submarines.

Did You Know?

Sonar technology, which uses sound waves in water, is not only used in marine navigation and fishing but also plays a vital role in various military applications, including underwater detection and navigation.

Under normal conditions of temperature and pressure, it is observed that sound waves propagate through air at an approximate velocity of 343 meters per second. However, when sound waves traverse water, their velocity is observed to be approximately 1,484 meters per second (Wei et al., 2022).

When sound waves come into contact with an obstruction, such as an object submerged in water, they experience a fundamental characteristic of wave behavior referred to as reflection. The emission of a sound wave originates from a transmitting source, propagates in an outward direction, encounters an object, undergoes reflection, and subsequently returns to the original sender. The receiver is responsible for detecting the reflected wave, commonly referred to as an echo, and conducting a comparison with the original wave's profile in order to ascertain the congruence of the reflected wave with the original wave. The distance to the object can subsequently be ascertained, given the established knowledge of the speed of sound in water. The wave propagates over a distance that is twice the round-trip distance to the object, and this propagation can be quantified in terms of time. The distance to an object in water can be determined by utilizing the formula: distance = (speed x time)/2, where the speed of sound in water remains constant.

The operation of sonar is based on the principle depicted in the diagram provided (Figure 7.6). The incident wave interacts with the object, undergoes reflection back towards the source, and subsequently reaches the receiver (Anitha et al., 2019).

The boat, which remains motionless, emits a sound wave that propagates via the water directly beneath its position on the lake's surface. The reception of the reflected sound wave occurs after a certain time interval. Given the known speed of sound in water, one can proceed to determine the depth of the lake. The velocity of sound in water remains consistent. The sound wave propagates in a linear trajectory, undergoes reflection upon encountering the lake bed, and subsequently comes back to the boat within a specific time interval. The calculation of the lake's depth involves dividing the entire travel time of the wave by two (Yang et al., 2020).

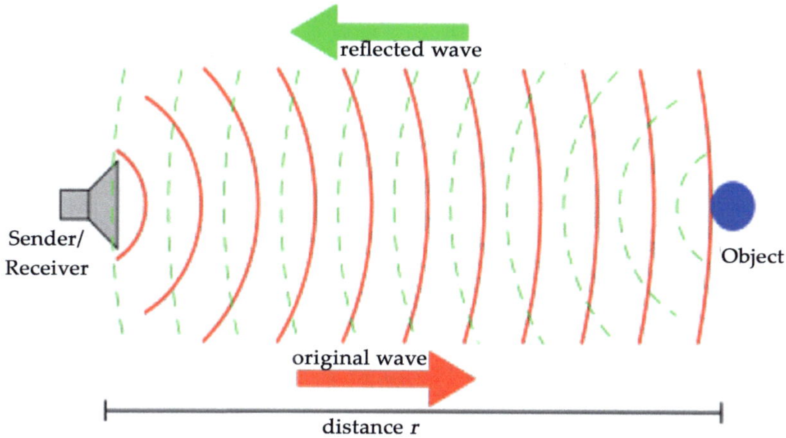

Figure 7.6. The underlying principle governing the functioning of sonar. The transmission of a wave occurs when a sender emits it, and subsequently, the wave is received following its reflection by an object (Source: Bartleby, Creative Commons License).

7.4.3. Ultrasound

Ultrasound waves are frequently employed in the field of medicine, particularly in the context of prenatal care. A notable application of ultrasound technology involves the visualization of a developing fetus within a pregnant woman, wherein an image is generated to depict the anatomical characteristics of the unborn child. Ultrasound waves function based on a comparable principle to sonar, employing the phenomenon of reflection. A transducer is employed for the purpose of producing and detecting sound waves with a frequency that exceeds the upper threshold of human auditory perception by several orders of magnitude (Cootney, 2001).

The acoustic waves experience partial reflection upon encountering the interfaces between distinct tissues within the human body. The transducer receives reflected waves and converts them into a digital signal, which is then visualized as an image. The waves interact with the diverse tissues and anatomical structures, causing them to scatter and attenuate in distinct manners. Consequently, an image depicting the desired structure, be it a tissue or an organ, is formed. The provided illustration serves as an exemplification of an ultrasound-generated image, a widely employed technology within the medical field (Figure 7.7). Ultrasound finds extensive application in diverse medical contexts, encompassing the monitoring of fetal growth, the identification and assessment of medical ailments, as well as the facilitation of medical interventions (Khismatullin & Akhatov, 2001).

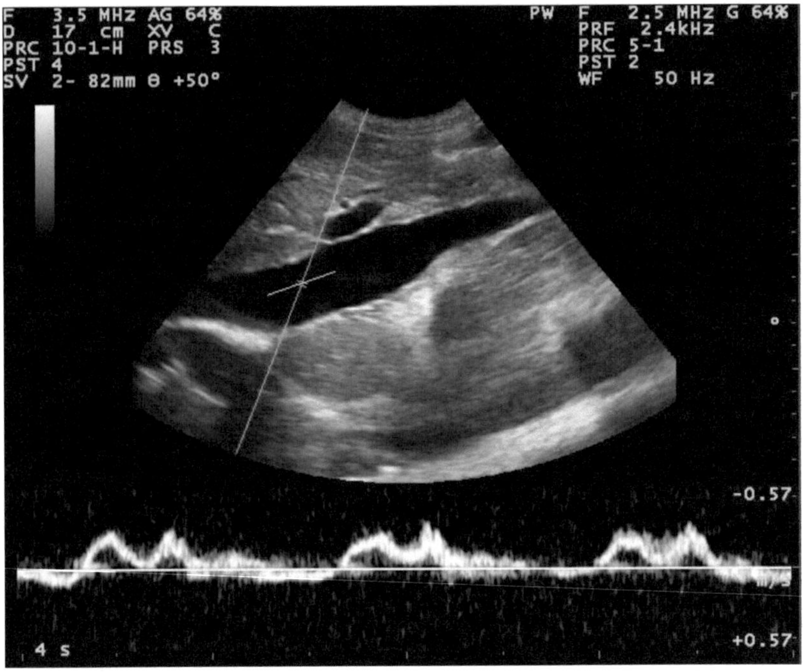

Figure 7.7. An ultrasound image of the inferior vena cava in a human (Source: Radiopaedia, Creative Commons License).

The partial reflection of ultrasound waves occurs due to various factors, such as refraction, scattering, and absorption, which can occur at the interface between different tissues.

Ultrasound waves with lower frequencies possess the ability to penetrate tissue to a greater extent, albeit at the cost of reduced image resolution. Ultrasound waves of higher frequency exhibit limited penetration depth, yet offer enhanced resolution for imaging purposes. The depth of the tissue boundary beneath the skin surface can be determined by employing the equation utilized for sonar. This equation incorporates variables such as the speed of the ultrasound wave, the total time considered the wave to be emitted and taken in by the transducer, as well as the depth of the tissue boundary (Gallo et al., 2018).

7.5. APPLICATIONS OF RADIO WAVES

Radio waves are a specific category of electromagnetic waves that reside within the electromagnetic spectrum, characterized by their extended wavelengths, which are estimated to be of considerable length. Radio waves are frequently employed for communication as well as detection purposes due to their capacity for long-distance travel. Two prominent examples of wave utilization include mobile phones as well as radar systems.

While radio waves have been shown to add to background radiation as well as are generally regarded as safe, it is advisable to maintain a considerable distance from the origin of radio wave emissions as a precautionary measure. The image provided depicts the internationally recognized hazard sign for harmful radio waves. This symbol serves to indicate that an individual is in proximity to a source emitting radio waves that may pose a potential risk. Consequently, it is advised for individuals to maintain a safe distance from the area in question.

It is imperative to acknowledge that although radio waves are usually thought of as safe, extended and high-intensity exposure to these waves may potentially result in adverse health consequences. Therefore, it is imperative to consistently exercise prudence when in close proximity to a potential emitter of deleterious radio frequencies (Hargreaves, 1969).

7.5.1. Mobile Phones

Radio waves possess the ability to traverse considerable distances and exhibit the capacity to penetrate solid structures, such as walls, thereby rendering them highly suitable for the transmission of cellular as well as telephone signals. Cellular phones are equipped with integrated receivers and transmitters. Transmitters function by transforming audio signals to electromagnetic radio waves that subsequently propagate over extensive distances and are directed and transmitted through cellular stations as well as towers until they ultimately reach the intended recipient's mobile device (Foster & Chou, 2014).

Remember

Mobile phones rely on radio waves for wireless communication. Understanding the different frequency bands used for mobile communication and the potential health implications of prolonged exposure to radio waves is important for users and policymakers.

One notable benefit of utilizing radio waves in this particular context is their ability to propagate through free space as electromagnetic waves, with a velocity of approximately 299,792 meters per second. This characteristic enables nearly instantaneous communication with minimal latency. The diagram presented below illustrates a conventional cellular tower, employed for the purpose of transmitting radio waves from the transmitter to the receiver. Typically, these structures exhibit considerable height in order to minimize wave absorption by the surrounding terrain, thereby facilitating enhanced communication quality.

It is noteworthy that although radio waves are thought of as safe, it is imperative to exercise responsible usage of cell phones and minimize prolonged as well as intense exposure to them. Furthermore, it is imperative to guarantee the appropriate maintenance and strategic placement of cell towers in order to mitigate any potential health hazards they may pose to individuals residing in close proximity (Sivani & Sudarsanam, 2012).

7.5.2. Radar

Radar, an acronym for radio detection as well as ranging, is an electromagnetic-based system utilized for the purpose of object detection within the atmosphere. The aforementioned principle bears resemblance to sonar, albeit with the substitution of sound waves with radio waves. This substitution is necessitated by the inadequate speed at which sound waves propagate through air, rendering them unsuitable for the purpose of detecting obstacles, like an approaching aircraft (Fujita et al., 2006).

Radio waves are propagated through the atmosphere by means of expansive antennas, undergo reflection upon encountering an obstruction, and subsequently travel back through the exact same medium to reach the receiver. The distance to the object can be determined by measuring the duration of the radio wave's round-trip time. The wave undergoes a complete round trip, covering a total distance of 2d, within a time interval of t. Given the constant velocity of electromagnetic waves in air, the distance d can be determined in the following manner (Nazari & Ghorbani, 2016):

$$d = ct/2$$

The provided visual representation depicts a radar antenna that possesses the capability to emit electromagnetic radio waves as well as subsequently detect the waves that have been reflected back. The dimensions of the radar system exhibit a direct correlation with the range of signal transmission and reception, thereby facilitating extensive communication and detection capabilities over considerable distances. It is noteworthy to mention that radar possesses numerous significant applications, including but not limited to air traffic management, weather monitoring, and military reconnaissance. Nevertheless, it is crucial to exercise responsible radar usage and guarantee that its implementation does not present any potential health hazards to individuals residing in close proximity (Kolanowska et al., 2018).

7.6. APPLICATION OF LIGHT WAVES

As previously stated, light waves are a type of electromagnetic wave that is situated within the visible portion of the electromagnetic spectrum. In contrast to radio waves, light rays are the sole category of electromagnetic waves perceptible to the human visual system.

Although the significance of light may often be overlooked, it possesses a multitude of practical applications within our everyday existence. Light waves are employed in the medical domain for diagnostic applications, including X-ray imaging and endoscopic procedures. Furthermore, these compounds find application in the management of specific medical ailments, such as their utilization in the use of photodynamic therapy for the treatment of cancer.

In addition to their utilization in medical contexts, light waves find extensive application across various technological domains, including fiber optic communication systems. These systems enable the rapid and efficient transmission of vast quantities of data over considerable distances. Light waves are employed in the fields of photography, film, as well as television to facilitate the acquisition and presentation of visual imagery (Saastamoinen, 1962).

It is imperative to acknowledge that light waves also possess a pivotal significance within our environment. Photosynthetic organisms rely on these energy sources to facilitate the process of photosynthesis, a vital mechanism for the sustenance of numerous ecosystems. Moreover, the presence of light pollution can exert detrimental effects on wildlife, causing disturbances in their migration patterns as well as breeding cycles.

In summary, it can be asserted that light waves, while being the sole form of electromagnetic waves perceptible to humans, possess a significance that transcends mere visual perception. Light waves are of utmost importance in various aspects of human existence, ranging from medical applications and communication systems to the preservation of ecosystems.

Photographers of a professional caliber frequently transport sizeable bags containing an assortment of equipment, among which the camera flash, a pivotal component of their toolkit, holds utmost significance. A camera flash is an instrument that emits a brief and concentrated burst of light in order to provide illumination for the purpose of capturing a photograph. The utilization of supplementary illumination through the flash has the potential to enhance the visual characteristics of the image, resulting in heightened clarity and enhanced definition (Bergé, 1998).

The flash produces a brief and intense emission of white light, effectively enhancing the brightness of the image and mitigating the impact of shadows or dim lighting circumstances. This holds particular significance in scenarios where the availability of natural light is inadequate, such as in the context of indoor or nighttime photography.

In addition to the field of photography, the utilization of light waves finds application in numerous other domains. Light waves are utilized in various medical imaging modalities, including X-rays, CT scans, as well as MRI scans. In addition, lasers find application in laser surgery, wherein a highly focused light beam is employed to eliminate or modify tissue.

Figure 7.8. The depicted visual representation showcases a luminous and intense emission of light characterized by a brilliant white hue, which is produced by a contemporary digital camera (Source: Shiken, Creative Commons License)

Light waves are utilized in the field of telecommunications, particularly in fiber optic systems for communication. These systems employ light pulses to efficiently transmit data over extensive distances with remarkable speed. Additionally, they find application in electronic displays, including computer monitors and televisions, wherein their function is to generate visual representations. In summary, the camera flash serves as a representative instance among the numerous utilizations of light waves in our everyday existence (Figure 7.8). Light waves are of utmost importance in various technologies and applications, spanning from photography to imaging in medicine, telecommunications to electronic displays (Dacey, 1962).

Activity 7.1

Explain the applications of microwaves in both communication and radar systems. Provide examples of how microwaves are utilized in modern technologies to enhance communication and facilitate radar-based operations.

SUMMARY

This chapter provides a comprehensive overview of the practical utilization of waves in various fields of technology and science. The chapter explores the importance of electromagnetic waves in communication technology, discussing radio waves, visible light waves, and infrared waves in communication systems, along with satellite communications. It further delves into the medical applications of waves, including ultrasound technology, X-rays, and gamma ray therapy, highlighting their significant role in diagnostics and treatments. The applications of sound waves, such as hearing in the human ear, sonar, and medical ultrasound, are also examined. Additionally, the chapter explores the uses of radio waves in mobile phones and radar systems, as well as the importance of light waves in optical communication and other applications. Overall, this chapter emphasizes the essential role of waves in shaping modern technology and medical advancements.

REVIEW QUESTIONS

1. What are the key communication technologies that utilize electromagnetic waves, and how do radio waves, visible light waves, and infrared waves play a role in communication systems?
2. Describe the medical applications of waves, including the use of ultrasound technology, X-rays in radiography, and gamma rays in medical therapy.
3. How are sound waves utilized in various applications, such as hear-

ing in the human ear, sonar, and medical ultrasound?

4. Explore the applications of radio waves, specifically in mobile phones and radar systems, and their significance in modern communication and navigation.
5. Discuss the importance of light waves in technology, particularly in optical communication and other applications.

REFERENCES

1. Anitha, U., Malarkkan, S., Jebaselvi, G. A., & Narmadha, R. (2019). Sonar image segmentation and quality assessment using prominent image processing techniques. *Applied Acoustics*, *148*, 300–307.
2. Bergé, L. (1998). Wave collapse in physics: principles and applications to light and plasma waves. *Physics reports*, *303*(5–6), 259–370.
3. Cella, U. M., Johnstone, R., & Shuley, N. (2009, November). Electromagnetic wave wireless communication in shallow water coastal environment: theoretical analysis and experimental results. In *Proceedings of the 4th International Workshop on Underwater Networks* (Vol. 2, pp. 1–8).
4. Cheng, J., Zhang, H., Ning, M., Raza, H., Zhang, D., Zheng, G., ... & Che, R. (2022). Emerging materials and designs for low-and multi-band electromagnetic wave absorbers: the search for dielectric and magnetic synergy?. *Advanced Functional Materials*, *32*(23), 2200123.
5. Choi, W., Jung, G., Kim, J., & Kwon, Y. (2009). Scalable small-signal modeling of RF CMOS FET based on 3-D EM-based extraction of parasitic effects and its application to millimeter-wave amplifier design. *IEEE Transactions on Microwave Theory and Techniques*, *57*(12), 3345–3353.
6. Cootney, R. W. (2001). Ultrasound imaging: principles and applications in rodent research. *Ilar Journal*, *42*(3), 233–247.
7. Dacey, G. C. (1962). Optical Masers in Science and Technology: Advances in the control of light waves give promise of important applications in science and technology. *Science*, *135*(3498), 71–74.
8. De Sanctis, M., Cianca, E., Araniti, G., Bisio, I., & Prasad, R. (2015). Satellite communications supporting internet of remote things. *IEEE Internet of Things Journal*, *3*(1), 113–123.
9. Foster, K. R., & Chou, C. K. (2014). Are children more exposed to radio frequency energy from mobile phones than adults?. *IEEE Access*, *2*, 1497–1509.
10. Fujita, S., Maeno, H., & Matsuoka, K. (2006). Radio-wave depolarization and scattering within ice sheets: a matrix-based model to link radar and ice-core measurements and its application. *Journal of Glaciology*, *52*(178), 407–424.
11. Gallo, M., Ferrara, L., & Naviglio, D. (2018). Application of ultrasound in food science and technology: A perspective. *Foods*, *7*(10), 164.

12. Hargreaves, J. K. (1969). Auroral absorption of HF radio waves in the ionosphere: A review of results from the first decade of riometry. *Proceedings of the IEEE*, *57*(8), 1348–1373.
13. Jeutter, D. C., & Josse, F. (1993). Design of a radio-linked implantable cochlear prosthesis using surface acoustic wave devices. *IEEE transactions on ultrasonics, ferroelectrics, and frequency control*, *40*(5), 469–477.
14. Khismatullin, D. B., & Akhatov, I. S. (2001). Sound–ultrasound interaction in bubbly fluids: Theory and possible applications. *Physics of Fluids*, *13*(12), 3582–3598.
15. Khodamoradi, E., Afrashi, S., Khoshgard, K., Fathi, F., Shahasavari, S., Azmoonfar, R., & Najafi, M. (2022). Simultaneous effect of gamma and Wi-Fi radiation on gamma-H2Ax expression in peripheral blood of rat: A radio-protection note. *Biochemistry and Biophysics Reports*, *30*, 101232.
16. Kodheli, O., Lagunas, E., Maturo, N., Sharma, S. K., Shankar, B., Montoya, J. F. M., ... & Goussetis, G. (2020). Satellite communications in the new space era: A survey and future challenges. *IEEE Communications Surveys & Tutorials*, *23*(1), 70–109.
17. Kolanowska, A., Janas, D., Herman, A. P., Jędrysiak, R. G., Giżewski, T., & Boncel, S. (2018). From blackness to invisibility–Carbon nanotubes role in the attenuation of and shielding from radio waves for stealth technology. *Carbon*, *126*, 31–52.
18. Maral, G., Bousquet, M., & Sun, Z. (2020). *Satellite Communications Systems: Systems, Techniques and Technology*. John Wiley & Sons.87(243), 6587–6589.
19. Moran, C. M., & Thomson, A. J. (2020). Preclinical ultrasound imaging—a review of techniques and imaging applications. *Frontiers in Physics*, *8*, 124.
20. Moridi, M. A., Kawamura, Y., Sharifzadeh, M., Chanda, E. K., & Jang, H. (2014). An investigation of underground monitoring and communication system based on radio waves attenuation using ZigBee. *Tunnelling and Underground Space Technology*, *43*, 362–369.
21. Nazari, M. E., & Ghorbani, A. (2016). Predicting a three-dimensional radar coverage area: Introducing a new method based on propagation of radio waves. *IEEE Antennas and Propagation Magazine*, *58*(1), 28–34.

22. Ratnam, V. V., Chen, H., Pawar, S., Zhang, B., Zhang, C. J., Kim, Y. J., ... & Yoon, S. R. (2020). FadeNet: Deep learning-based mm-wave large-scale channel fading prediction and its applications. *IEEE Access*, *9*, 3278–3290.
23. Raza, U., & Salam, A. (2020). Wireless underground communications in sewer and stormwater overflow monitoring: radio waves through soil and asphalt medium. *Information*, *11*(2), 98.
24. Saastamoinen, J. (1962). The effect of path curvature of light waves on the refractive index application to electronic distance measurement. *The Canadian Surveyor*, *16*(2), 98–100.
25. Sivani, S., & Sudarsanam, D. (2012). Impacts of radio-frequency electromagnetic field (RF-EMF) from cell phone towers and wireless devices on biosystem and ecosystem-a review. *Biology and Medicine*, *4*(4), 202.
26. Topfer, F., & Oberhammer, J. (2015). Millimeter-wave tissue diagnosis: The most promising fields for medical applications. *IEEE Microwave Magazine*, *16*(4), 97–113.
27. Underwood, J. H., & Barbee, T. W. (1981). Layered synthetic microstructures as Bragg diffractors for X rays and extreme ultraviolet: theory and predicted performance. *Applied Optics*, *20*(17), 3027–3034.
28. Waksman, R., White, R. L., Chan, R. C., Bass, B. G., Geirlach, L., Mintz, G. S., ... & Leon, M. B. (2000). Intracoronary γ-radiation therapy after angioplasty inhibits recurrence in patients with in-stent restenosis. *Circulation*, *101*(18), 2165–2171.
29. Wang, G., Ong, S. J. H., Zhao, Y., Xu, Z. J., & Ji, G. (2020). Integrated multifunctional macrostructures for electromagnetic wave absorption and shielding. *Journal of Materials Chemistry A*, *8*(46), 24368–24387.
30. Wei, Y., Duan, Y., & An, D. (2022). Monitoring fish using imaging sonar: Capacity, challenges and future perspective. *Fish and Fisheries*, *23*(6), 1347–1370.
31. Wu, S., Wang, H., & Youn, C. H. (2014). Visible light communications for 5G wireless networking systems: from fixed to mobile communications. *Ieee Network*, *28*(6), 41–45.
32. Yang, H., Lee, K., Choo, Y., & Kim, K. (2020). Underwater acoustic research trends with machine learning: Passive SONAR applications. *Journal of Ocean Engineering and Technology*, *34*(3), 227–236.

33. Yoneyama, M., Fujimoto, J. I., Kawamo, Y., & Sasabe, S. (1983). The audio spotlight: An application of nonlinear interaction of sound waves to a new type of loudspeaker design. *The Journal of the Acoustical Society of America*, *73*(5), 1532–1536.
34. Zhang, B., Chen, W., Wang, P., Dai, S., Li, H., Lu, H., ... & Yang, J. (2019). Particle swarm optimized polarization beam splitter using metasurface-assisted silicon nitride Y-junction for mid-infrared wavelengths. *Optics Communications*, *451*, 186–191.

CHAPTER 8

Future of Waves

Contents

8.1. INTRODUCTION

Waves, in all their forms, have been integral to the progress of modern society, shaping how we communicate, explore, and understand the world. From the familiar electromagnetic waves used in wireless communication to the seismic waves that reveal the Earth's structure, waves are fundamental to our daily lives. The significance of waves in modern society lies in their ability to transmit information, energy, and insights across vast distances, revolutionizing various industries and enhancing our quality of life (Guerreiro et al., 2018).

One of the most prominent areas where waves play a crucial role is communication technology. Electromagnetic waves, such as radio waves and microwaves, underpin the global telecommunication network, enabling seamless voice and data transmission across continents. The evolution of wave technology has led to faster and more reliable communication, empowering businesses, connecting communities, and fostering global collaboration. As the demand for data-intensive applications grows, the significance of waves in facilitating high-speed internet and future technologies cannot be overstated (Zacharias et al., 2014).

Beyond communication, waves have transformed the landscape of medical science and healthcare. Ultrasound technology, which uses sound waves, revolutionized diagnostic imaging, enabling non-invasive visualization of internal organs and tissues. Waves in the form of X-rays and gamma rays have become indispensable tools for medical imaging and cancer treatment. With ongoing advancements, waves are paving the way for targeted therapies, precision medicine, and innovative diagnostic techniques that promise to revolutionize healthcare and improve patient outcomes (Molina et al., 2020).

8.2. THE EVOLVING LANDSCAPE OF WAVE TECHNOLOGY

The landscape of wave technology is undergoing rapid transformation, driven by relentless research, technological innovations, and the quest for sustainability. As we venture into the future, several key trends are shaping the evolution of wave technology.

One of the most promising developments is the advancement of communication technology. With the rise of the Internet of Things (IoT), autonomous vehicles, and smart cities, there is a growing demand for ultra-reliable, low-latency communication networks. The emergence of 5G technology has laid the foundation for unprecedented data speeds and connectivity, but the future holds even more potential. The era of 6G and beyond promises to push the boundaries of wireless communication, offering data rates in the terabits per second, empowering real-time applications, and enabling new paradigms of human-machine interaction (Thongpapanl, 2012).

Quantum waves represent another transformative area of exploration. Quantum computing, leveraging the principles of quantum mechanics, holds the promise of solving complex problems exponentially faster than classical computers. Quantum communication offers unparalleled security through quantum encryption, protecting data from potential cyber threats. As research progresses, harnessing the power of quantum waves may unlock groundbreaking discoveries in fields like drug development, material science, and cryptography (Bertozzi et al., 2015).

Additionally, nanowaves and miniaturization are shaping the future of technology. Nanotechnology enables the manipulation of materials at the atomic or molecular level, leading to the development of nanowave devices with extraordinary capabilities. These miniature devices find applications in medicine, environmental sensing, and even quantum computing. The miniaturization of wave technology opens new frontiers, allowing for wearable devices, implantable sensors, and energy-efficient electronics.

Furthermore, waves are poised to revolutionize sustainable energy production. Wave energy conversion and ocean thermal energy conversion (OTEC) harness the power of ocean waves to generate electricity. Advances in solar and wind energy, both of which rely on wave principles, offer renewable and clean alternatives to traditional energy sources, contributing to a more sustainable future (Piri, 2018).

8.3. ADVANCEMENTS IN COMMUNICATION TECHNOLOGY

In the rapidly evolving landscape of communication technology, the pursuit of faster, more reliable, and efficient wireless communication has led to significant advancements. Next-generation wireless communication aims to address the ever-increasing demand for data-intensive applications, connect an expanding array of devices in the Internet of Things (IoT), and enable seamless connectivity in an increasingly connected world (Przybylski &

Weinstein, 2013). The fifth-generation (5G) wireless communication system, which has seen significant deployment globally, marked a transformative step forward. 5G offers significantly higher data speeds, lower latency, and enhanced network capacity compared to its predecessor, 4G LTE. With data rates exceeding several gigabits per second, 5G enables real-time applications, ultra-HD video streaming, and virtual and augmented reality experiences. Moreover, its low latency is vital for critical applications like autonomous vehicles, remote surgery, and industrial automation (Efobi et al., 2018).

8.3.1. Beyond 5G: The Era of 6G and Beyond

While 5G continues to roll out, the quest for even more advanced wireless communication is already underway. The era of 6G and beyond promises to push the boundaries of wireless connectivity to unimaginable heights. Research and development efforts are exploring several key areas to shape the future of wireless communication.

Tip

The future of communication technology lies in the development of 6G and beyond, which promises even faster data rates, ultra-low latency, and unprecedented connectivity. Stay updated on the latest advancements in this area to harness its potential in various industries.

- ***Increased Data Rates***: 6G aims to achieve data rates in the order of terabits per second (Tbps), providing seamless multi-gigabit connectivity to users. This will facilitate instant downloads, ultra-high-definition streaming, and immersive virtual experiences that will redefine human-machine interaction (Ghosh et al., 2014).
- ***Ultra-Low Latency***: 6G is set to introduce ultra-low latency, reducing communication delays to a few microseconds. This will enable near-instantaneous response times, making applications like real-time gaming, telemedicine, and remote industrial control even more efficient and reliable.
- ***Massive Connectivity***: With the proliferation of IoT devices, 6G seeks to support an unprecedented number of connected devices per square kilometer. This massive connectivity will pave the way for smart cities, smart transportation, and seamless integration of IoT devices in various industries (Liu et al., 2018).

- ***Terahertz Spectrum*:** The use of terahertz frequencies is a key aspect of 6G research. Terahertz waves offer significantly higher bandwidths, enabling data transmission at extraordinary speeds. However, exploiting this spectrum comes with technical challenges, including signal attenuation and propagation limitations.
- ***AI-Enabled Communication*:** Artificial intelligence (AI) will play a pivotal role in 6G networks, optimizing resource allocation, managing network traffic, and enhancing security (Figure 8.1). AI-driven communication will make networks more adaptive, efficient, and capable of dynamically meeting user demands (Yan et al., 2020).

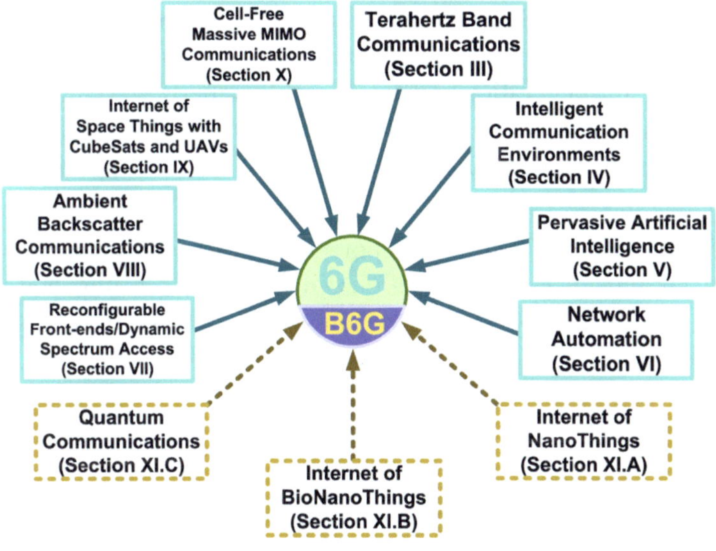

Figure 8.1. The envisioned key enabling technologies for 6G and beyond wireless communications systems (Source: Ahan Kak, Creative Commons License).

8.3.2. Terahertz Communication and Its Potential

One of the most exciting prospects in the realm of next-generation communication is terahertz communication. Terahertz waves occupy the frequency range between microwave and infrared waves, spanning from 0.1 to 10 terahertz (THz). This spectrum holds enormous potential for high-speed wireless communication and data transmission. Terahertz waves offer bandwidths that are orders of magnitude larger than current microwave and millimeter-wave bands used in 5G. This vast bandwidth can accommodate a colossal amount of data, paving the way for blazing-fast wireless connectivity. Terahertz communication could offer data rates of several terabits per second, revolutionizing internet access and information exchange (Koch, 2007).

However, terahertz communication faces numerous technical challenges. Terahertz waves have relatively short wavelengths and are easily absorbed by water vapor and atmospheric gases. This absorption leads to significant signal attenuation, limiting the transmission distance and requiring sophisticated signal processing and beamforming techniques to mitigate signal loss.

Despite these challenges, researchers and engineers are making strides in terahertz communication. Advancements in terahertz source technology, such as quantum cascade lasers and terahertz quantum well photodetectors, have improved the generation and detection of terahertz waves. Additionally, advancements in metamaterials and nanostructured devices offer promise in manipulating and guiding terahertz waves, enabling better propagation and communication.

Terahertz communication holds potential not only for ultra-fast wireless data transfer but also for applications in imaging, sensing, and security screening. Terahertz waves can penetrate certain materials, such as fabrics and plastics, without the harmful ionizing effects of X-rays, making them suitable for non-invasive imaging and security screening applications (Elayan et al., 2019).

8.4. QUANTUM WAVES AND INFORMATION PROCESSING

Quantum waves form the foundation of quantum mechanics, a branch of physics that describes the behavior of particles at the smallest scales of matter and energy. In the quantum realm, particles such as electrons and photons exhibit both particle-like and wave-like characteristics, known as wave-particle duality. Quantum waves are described by mathematical functions known as wavefunctions, which determine the probability of finding a particle in a particular state or location (Figure 8.2).

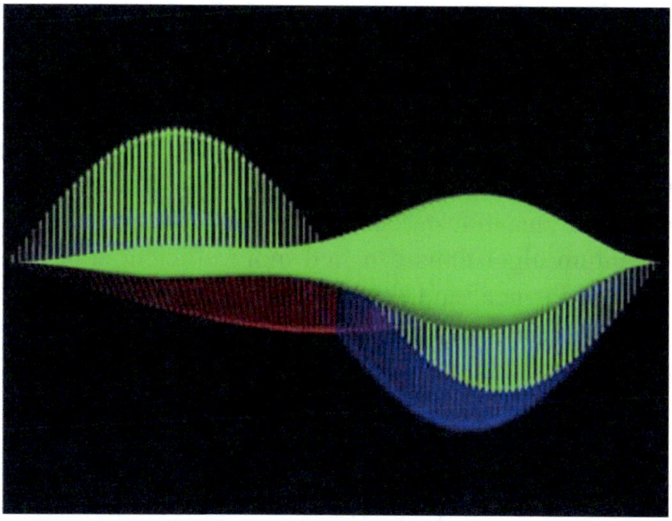

Figure 8.2. Quantum waves in 3D (Source: Steve Spicklemire, Creative Commons License).

One of the most striking features of quantum waves is superposition, wherein particles can exist in multiple states simultaneously. This concept is in stark contrast to classical physics, where particles exist in definite states. Additionally, quantum waves can be entangled, where the state of one particle is directly correlated to the state of another, regardless of the distance between them. These properties underpin the incredible potential of quantum waves for information processing and communication (Curtis & Hurtak, 2004).

8.4.1. Quantum Computing and its Applications

Quantum computing harnesses the unique properties of quantum waves to perform computational tasks in a fundamentally different way than classical computers. Traditional computers use bits, which represent information as either 0 or 1, in a series of binary digits. Quantum computers, on the other hand, use quantum bits or qubits, which can exist in a superposition of states, allowing them to represent and process multiple pieces of information simultaneously.

Quantum computing's immense computational power arises from its ability to explore multiple solutions simultaneously through superposition and entanglement. This allows quantum computers to solve certain problems exponentially faster than classical computers. One of the most famous quantum algorithms is Shor's algorithm, which efficiently factors large numbers, posing a significant threat to modern encryption methods (Spector et al., 1999).

Quantum computing holds great promise for a wide range of applications. In cryptography, it can break current cryptographic schemes and motivate the development of quantum-resistant encryption methods. In optimization problems, quantum algorithms can find more efficient solutions in fields such as logistics, finance, and drug discovery. Quantum simulations can model complex systems, such as quantum chemistry and materials science, with greater accuracy than classical methods.

However, building and maintaining stable quantum computers are significant challenges. Quantum systems are highly sensitive to environmental disturbances, which can cause errors and disrupt quantum states. Researchers are actively developing error-correcting codes and fault-tolerant techniques to address these challenges and pave the way for scalable and reliable quantum computing (Perdomo et al., 2008).

8.4.2. Quantum Communication and Cryptography

Quantum communication utilizes the principles of quantum mechanics to achieve secure and unbreakable information transfer. Quantum cryptography, in particular, offers a quantum-safe alternative to traditional cryptographic methods vulnerable to quantum attacks.

Quantum key distribution (QKD) is a fundamental application of quantum communication. QKD enables two parties to establish a secret key using quantum properties, such as entanglement and the no-cloning theorem. The key is then used for secure communication through traditional cryptographic protocols. The remarkable feature of QKD is that any attempt to eavesdrop or intercept the quantum information disturbs the quantum states, immediately revealing the presence of an attacker. This property ensures unconditional security, making it impossible for an adversary to gain access to the secret key without detection (Pirandola et al., 2008).

Quantum communication also plays a role in secure quantum teleportation and quantum teleportation-based quantum cryptography, where quantum information is transmitted instantaneously, without traveling through the intervening space. While not suitable for transmitting classical information, quantum teleportation holds significance in quantum computing and quantum communication protocols.

The implementation of quantum communication networks faces challenges related to the preservation of quantum states over long distances and in noisy environments. Quantum repeaters, which can extend the range of quantum communication, and quantum error correction methods are actively researched to overcome these challenges and make quantum communication a practical reality (Ng et al., 2020).

8.5. NANOWAVES AND MINIATURIZATION

Nanotechnology, the study and manipulation of matter at the nanoscale, has had a profound impact on various scientific disciplines, including the field of waves and wave technology. Nanowaves refer to waves that interact with nanoscale structures, either as a source or during propagation, opening up exciting possibilities for miniaturization, enhanced functionality, and novel applications.

In nanotechnology, researchers can engineer materials with precise control over their physical and chemical properties at the nanoscale. This level of control allows for tailoring materials to interact with waves in unique ways, enabling the creation of nanowave devices with unprecedented capabilities. By designing materials with specific electromagnetic properties, nanotechnology has paved the way for wave manipulation and control, leading to innovations in optics, electronics, and communication (Cao et al., 2021).

8.5.1. Nanowave Devices and Applications

Nanowave devices encompass a diverse range of technologies that leverage nanomaterials and nanostructures to manipulate and control waves. Some notable nanowave devices and their applications include:

- ***Metamaterials***: Metamaterials are engineered composites with properties not found in natural materials. They possess negative refractive indices, allowing for the creation of “invisibility cloaks” that redirect light around an object, rendering it invisible. Metamaterials find applications in lens design, imaging, and stealth technology.
- ***Nanophotonics***: Nanophotonics explores the interaction of light with nanoscale structures. Plasmonic nanoparticles, for example, can confine light to subwavelength scales, enabling high-resolution imaging and sensing. Nanophotonics holds promise

in developing efficient and compact photonic devices for data communication and computing (Zimmermann et al., 2008).

- ***Nanoelectronics*:** Nanowires and nanotubes offer unique electronic properties due to their size and quantum effects. These nanostructures can manipulate electrical signals at the nanoscale, leading to advancements in computing, memory devices, and sensors.
- ***Nanoresonators*:** Nanoscale resonators can vibrate at specific frequencies, making them valuable in sensing applications. Nanoresonators enable highly sensitive and selective detection of chemical and biological molecules, with potential applications in medical diagnostics and environmental monitoring.
- ***Nanoantennas*:** Nanoantennas interact with electromagnetic waves in the optical range, enabling enhanced light-matter interactions. They have applications in photodetection, solar energy harvesting, and on-chip optical communication (Hackett et al., 2018).

8.5.2. Challenges and Opportunities in Miniaturization

While nanowaves and miniaturization offer tremendous opportunities for technological advancements, they also present several challenges that need to be addressed:

- ***Fabrication Techniques*:** Creating precise nanoscale structures with consistency and cost-effectiveness is a significant challenge. Advanced fabrication techniques, such as electron beam lithography and nanoimprint lithography, are essential for realizing nanowave devices (Magisetty et al., 2018).
- ***Losses and Noise*:** At the nanoscale, materials can exhibit higher losses and increased noise levels, affecting the efficiency and reliability of nanowave devices. Researchers must address these issues to optimize device performance.
- ***Quantum Effects*:** In the realm of nanowaves, quantum effects can become dominant, influencing wave-matter interactions. Understanding and controlling these effects is essential for developing robust nanowave technologies.
- ***Integration and Scalability*:** Integrating nanowave devices into practical systems and scaling them for commercial applications is

a complex task. Ensuring compatibility with existing technologies and developing scalable manufacturing processes are vital for widespread adoption.

Despite the challenges, miniaturization offers promising opportunities:

- ***Compact Devices***: Miniaturization allows for the development of compact devices with reduced size, weight, and power consumption. This has significant implications for portable electronics, wearable technology, and space-constrained applications.
- ***Enhanced Functionality***: Nanowave devices offer enhanced functionality due to their ability to manipulate waves at the nanoscale. This opens up possibilities for improved imaging, sensing, and data processing capabilities.
- ***Quantum Technologies***: Miniaturization has fueled advancements in quantum technologies, including quantum computing and quantum communication. Nanoscale components play a crucial role in the development of quantum devices and systems.
- ***Energy Efficiency***: Miniaturization can lead to energy-efficient devices, conserving resources and reducing environmental impact in various applications, from electronics to renewable energy systems (Ghanbari et al., 2017).

8.6. HARNESSING WAVES FOR SUSTAINABLE ENERGY

Following are the ways by which the waves can be harnessed for sustainable energy:

8.6.1. Wave Energy Conversion and Power Generation

Wave energy conversion is a promising avenue for sustainable energy generation, leveraging the kinetic energy of ocean waves to produce electricity. As waves propagate across the ocean's surface, they carry significant energy potential, making them an abundant and renewable resource.

Various technologies have been developed to harness wave energy and convert it into usable power. One approach is through the use of wave energy converters (WECs), which come in different forms such as point absorbers, oscillating water columns, and attenuators. Point absorbers use floating buoys tethered to the seabed, converting the up-and-down motion of waves into mechanical energy that drives electricity-generating turbines. Oscillating water columns, on the other hand, use the rising and falling of waves within a chamber to generate air pressure, which drives a turbine connected to an electrical generator. Attenuators, often placed perpendicular to wave direction, capture energy from the relative motion between the buoy and the waves (Figure 8.3).

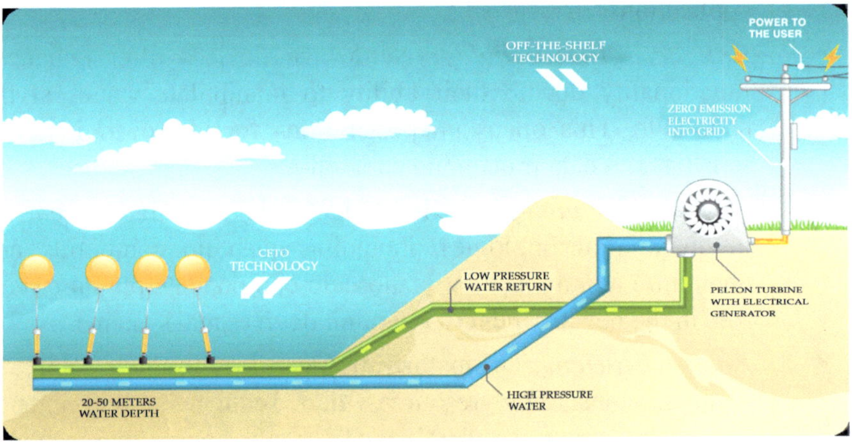

Figure 8.3. Wave Power Diagram (Source: The Earth Project, Creative Commons License).

Wave energy conversion offers several advantages, including predictability and high energy density. Unlike solar and wind energy, wave energy is relatively predictable and less influenced by weather conditions, making it a stable and consistent source of power. Additionally, waves carry a higher energy density compared to wind, enabling the generation of significant electricity from a relatively small area of the ocean.

Despite its potential, wave energy conversion faces challenges in terms of technological complexity, installation costs, and environmental impact. Designing efficient and robust WECs that can withstand harsh ocean conditions while being cost-effective remains a key challenge. Moreover, the deployment of wave energy devices requires careful consideration of their environmental effects on marine ecosystems and coastal regions (Wang et al., 2020).

8.6.2. Ocean Thermal Energy Conversion (OTEC)

Ocean Thermal Energy Conversion (OTEC) is another innovative method of harnessing the ocean's energy for sustainable power generation. OTEC takes advantage of the temperature difference between warm surface waters and cold deep waters in tropical regions to produce electricity.

The process involves using a closed-loop or open-loop system. In a closed-loop system, warm surface water is used to vaporize a working fluid, such as ammonia, which drives a turbine connected to an electrical generator. The vapor is then condensed using cold deep water to complete the cycle. In an open-loop system, warm surface water is directly used to boil a low-boiling-point fluid, producing steam to drive the turbine, while cold water from the ocean serves as a coolant.

OTEC offers several advantages, including a constant and reliable source of energy. Unlike solar and wind energy, which are intermittent, the temperature difference between surface and deep waters remains relatively constant throughout the year, providing a continuous power source. Additionally, OTEC has the potential for providing desalinated water as a byproduct, addressing water scarcity in coastal regions.

Challenges facing OTEC include the need for large temperature differences between surface and deep waters and the distance between potential OTEC sites and areas of energy demand. Implementing OTEC on a large scale would require significant infrastructural investments and careful consideration of its environmental impact, including the potential for affecting marine ecosystems (Rajagopalan & Nihous, 2013).

8.6.3. Advances in Solar and Wind Energy

Solar and wind energy continue to be at the forefront of sustainable energy solutions, with ongoing advancements in technology and implementation. Both solar and wind energy offer abundant and clean sources of power, making them crucial components of the renewable energy mix.

- ***Solar Energy***: Advances in solar photovoltaic (PV) technology have resulted in increased efficiency and reduced costs of solar panels. Innovations such as thin-film solar cells, perovskite solar cells, and bifacial solar panels have expanded the range of applications and improved energy conversion rates. Additionally, advancements in solar energy storage solutions, such as batteries and other energy storage systems, are enhancing the ability to store excess solar power for use during cloudy periods or at night.

- ***Wind Energy***: Wind energy continues to make significant progress in both onshore and offshore applications. Improved turbine designs, taller towers, and longer blades have led to increased energy capture and reduced costs of wind energy production. Offshore wind farms, in particular, have gained traction due to the availability of stronger and more consistent winds at sea, allowing for the development of large-scale wind energy projects closer to population centers.

Integration of solar and wind energy with energy storage technologies and smart grid solutions is enabling a more reliable and stable renewable energy supply. Combining the intermittent nature of solar and wind with effective energy storage systems ensures continuous power availability and grid stability, making these technologies more viable for widespread implementation.

In conclusion, harnessing waves for sustainable energy presents diverse opportunities to tap into the vast energy potential of the ocean. Wave energy conversion and OTEC offer promising solutions to generate clean electricity from the natural movements and temperature variations of ocean waters. Alongside these wave-based technologies, continued advances in solar and wind energy are contributing to a more diversified and sustainable energy mix. As these technologies mature and become more economically competitive, they will play an increasingly significant role in reducing greenhouse gas emissions, mitigating climate change, and building a more sustainable energy future (Neupane et al., 2022).

8.7. WAVES IN MEDICINE AND HEALTHCARE

Waves have revolutionized the field of medical imaging, allowing healthcare professionals to visualize internal structures and diagnose medical conditions non-invasively. Various imaging modalities utilize different types of waves to provide detailed images of the human body.

8.7.1. Non-Invasive Therapies and Treatments

Waves have also contributed to the development of non-invasive therapies and treatments, offering patients safer and more comfortable medical interventions (Figure 8.4).

- ***Extracorporeal Shockwave Therapy (ESWT):*** ESWT utilizes high-energy shockwaves to treat various musculoskeletal conditions, such as plantar fasciitis, tendinopathies, and calcific shoulder tendinopathy. The shockwaves stimulate healing and tissue repair, reducing the need for invasive procedures.

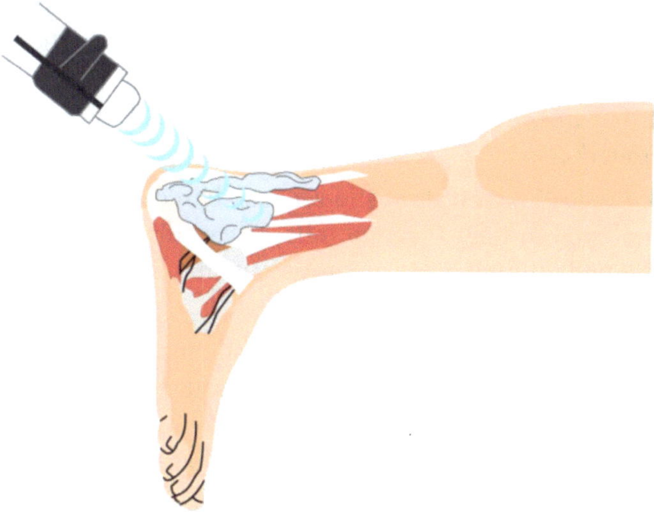

Figure 8.4. Schematic of the Shockwave Therapy (Source: Pain Relief Specialist, Creative Commons License).

- ***High-Intensity Focused Ultrasound (HIFU):*** HIFU focuses ultrasound waves to create heat at precise locations within the body. This non-invasive technique is used for thermal ablation of tumors and has shown promising results in treating prostate cancer, uterine fibroids, and liver tumors.
- ***Transcranial Magnetic Stimulation (TMS):*** TMS uses magnetic fields to stimulate specific regions of the brain non-invasively. It is employed in the treatment of depression, migraines, and certain neurological disorders, offering a targeted and well-tolerated alternative to traditional treatments (Loo & Mitchell, 2005).

Remember

Waves in medicine, such as ultrasound and electromagnetic waves, are increasingly being used for non-invasive therapies and treatments. Understanding the principles and applications of these wave-based medical technologies can lead to groundbreaking advancements in healthcare.

8.7.2. Wave-Based Diagnostics and Sensors

Waves have enabled the development of diagnostic tools and sensors that provide valuable information for disease detection and monitoring.

- ***Biosensors*:** Biosensors utilize waves, such as optical waves or electrical signals, to detect and measure specific biological markers in body fluids. They are used for glucose monitoring in diabetes, detecting infectious agents, and identifying cancer biomarkers.
- ***Acoustic Sensors*:** Acoustic sensors, based on sound waves, are employed in lung function tests, where they measure the speed of sound through the respiratory system to assess lung health and function.
- ***Continuous Wave Doppler*:** Continuous wave Doppler is a specialized ultrasound technique used to assess blood flow through vessels and the heart. It aids in diagnosing conditions like heart valve abnormalities, blood clots, and arterial stenosis.

These advancements in wave-based diagnostics and sensors have enabled earlier and more accurate disease detection, leading to timely interventions and improved patient outcomes (Kim et al., 2015).

8.8. SPACE EXPLORATION AND ASTRONOMY

The following are the ways in which waves can be used in space exploration.

8.8.1. Wave Technology in Space Telescopes

Wave technology plays a crucial role in space exploration and astronomy, particularly in the design and operation of space telescopes. Space telescopes utilize various types of waves to observe distant celestial objects and phenomena beyond the reach of Earth-based observatories.

- ***Optical and Infrared Telescopes*:** Optical and infrared telescopes use electromagnetic waves in the visible and infrared regions of the spectrum to capture images and collect data from distant stars, galaxies, and nebulae. They offer valuable insights into the composition, temperature, and movement of celestial objects.
- ***Radio Telescopes*:** Radio telescopes detect and analyze radio waves emitted by celestial sources. They are vital in studying radio galaxies, pulsars, and the cosmic microwave background

radiation, providing essential information about the early universe.

- ***X-ray and Gamma-ray Telescopes*:** X-ray and gamma-ray telescopes observe high-energy waves from sources such as black holes, neutron stars, and supernovae. These telescopes are crucial for studying extreme cosmic phenomena and understanding the most energetic processes in the universe (Ando et al., 2021).

8.8.2. Probing the Universe with Advanced Waves

Advanced wave technology has enabled the development of cutting-edge instruments and observatories that can probe the universe in unprecedented detail.

- ***Gravitational Wave Detectors*:** Gravitational wave detectors, such as LIGO (Laser Interferometer Gravitational-Wave Observatory), are designed to detect and measure tiny ripples in the fabric of spacetime caused by cataclysmic events, like merging black holes or neutron stars. These detections provide a new way to explore the cosmos and study phenomena beyond the reach of traditional electromagnetic telescopes.
- ***Space Interferometry*:** Interferometry involves combining signals from multiple telescopes to create a virtual telescope with much higher resolution than individual telescopes. Space-based interferometers, like the Hubble Space Telescope and the upcoming James Webb Space Telescope, allow astronomers to observe distant objects with unprecedented clarity, enabling groundbreaking discoveries about the universe's origins and evolution.

Did You Know?

Advanced wave technologies play a crucial role in probing the universe and studying cosmic phenomena. Gravitational waves, for instance, provide valuable insights into black holes, neutron stars, and the origins of the universe.

8.8.3. Future Missions and Discoveries

The future of space exploration and astronomy holds exciting prospects for uncovering the mysteries of the universe using advanced wave technology.

- ***James Webb Space Telescope (JWST):*** The JWST, set to launch, is poised to be the most powerful space telescope ever built. Operating primarily in the infrared range, JWST will reveal distant galaxies, study exoplanets, and shed light on the early universe.
- ***Wide Field Infrared Survey Telescope (WFIRST):*** WFIRST will conduct wide-field imaging and surveys, studying dark energy, exoplanets, and galaxies, among other astronomical phenomena. It will enhance our understanding of the universe's expansion and evolution.
- ***Gravitational Wave Astronomy:*** Gravitational wave observatories will continue to expand, detecting more gravitational wave events and unveiling new aspects of the universe. The inclusion of additional detectors worldwide will improve localization of gravitational wave sources and open new possibilities for multi-messenger astronomy.
- ***Radio Astronomy Advancements:*** Advancements in radio telescopes, such as the Square Kilometer Array (SKA), will enable astronomers to survey the sky with unprecedented sensitivity, unveiling previously unknown radio sources and enhancing our understanding of cosmic magnetism and the early universe (Kellermann & Moran, 2001).

8.9. ENVIRONMENTAL MONITORING AND CLIMATE STUDIES

Following are the ways in which waves can be used in Environmental Monitoring and Climate Studies:

8.9.1. Wave-Based Remote Sensing for Earth Observation

Wave-based remote sensing plays a pivotal role in environmental monitoring and climate studies, enabling scientists to gather critical data about Earth's surface, atmosphere, and oceans from space-based instruments.

- ***Satellite Imaging***: Satellites equipped with optical and infrared sensors capture high-resolution images of the Earth's surface, providing valuable information about land cover, vegetation health, and changes in urban areas. These images are essential for monitoring deforestation, urbanization, and natural disasters.

- ***Synthetic Aperture Radar (SAR)***: SAR uses microwave signals to penetrate cloud cover and collect data about the Earth's surface. SAR imagery is particularly useful for monitoring changes in polar ice caps, detecting land subsidence, and mapping ocean currents and oil spills.
- ***LiDAR (Light Detection and Ranging)***: LiDAR sensors on satellites and aircraft measure the distance between the sensor and the Earth's surface by emitting and detecting laser pulses. LiDAR data enable precise topographic mapping, forest inventory, and coastal erosion studies (Villa et al., 2021).

8.9.2. Monitoring Climate Change with Waves

Wave-based remote sensing is instrumental in monitoring climate change and its impact on the Earth's ecosystems and climate systems.

- ***Sea Level Rise***: Satellite altimeters use radar pulses to measure variations in sea surface height, providing crucial data on global sea level rise. This information helps assess the impact of climate change on coastal regions, track ocean currents, and study changes in polar ice sheets.
- ***Ocean Temperature and Heat Content:*** Ocean temperature measurements obtained through infrared and microwave radiometry help monitor variations in sea surface temperature and heat content. Understanding these changes is vital for assessing ocean circulation patterns and their influence on climate.
- ***Atmospheric Monitoring***: Remote sensing instruments aboard satellites measure various atmospheric parameters, including temperature, humidity, and greenhouse gas concentrations. This data is crucial for studying climate patterns, assessing air quality, and monitoring the ozone layer (Fox & Alder, 1989).

8.9.3. Spaceborne Sensors and Global Impact

Spaceborne sensors and remote sensing data have a significant global impact on environmental monitoring and climate studies.

- ***Global Coverage***: Satellites with wave-based sensors provide near-global coverage, enabling researchers to study remote and inaccessible regions of the Earth. This comprehensive data is essential for understanding global climate patterns and ecosystem dynamics.

- ***Early Warning Systems*:** Remote sensing data helps in developing early warning systems for natural disasters such as hurricanes, wildfires, and tsunamis. Timely information allows for better preparedness and disaster response, minimizing the impact on human lives and infrastructure.
- ***Climate Modeling and Predictions*:** Remote sensing data feeds into climate models, providing essential inputs for predicting future climate scenarios and understanding the potential consequences of climate change. This information is vital for policy decisions and climate change mitigation strategies.
- ***Biodiversity and Conservation*:** Satellite data aids in monitoring changes in biodiversity, identifying habitat loss, and studying the distribution of endangered species. This information contributes to conservation efforts and ecosystem management (Mihók et al., 2017).

Activity 8.1

Explain the concept of Terahertz communication and its potential applications in various fields, including wireless communication, medical imaging, and security screening. Also, discuss the challenges and limitations associated with harnessing Terahertz waves for practical use.

SUMMARY

The chapter titled "The Future of Waves" explores the significance of waves in modern society and the evolving landscape of wave technology. It delves into cutting-edge advancements in communication technology, quantum waves, and nanowaves. Additionally, the chapter discusses the use of waves for sustainable energy generation, their applications in medicine and healthcare, and their role in space exploration and astronomy. Furthermore, it delves into wave-based remote sensing for environmental monitoring and climate studies. Overall, the chapter provides a comprehensive look into the future prospects and potential impact of waves in various fields, highlighting their transformative role in shaping our world.

REVIEW QUESTIONS

1. What are some of the advancements in communication technology discussed in this chapter, and what is the potential impact of 6G and terahertz communication?
2. Explore the applications of quantum waves in information processing. How does quantum computing differ from classical computing, and what are its potential applications?
3. Describe the concept of nanowaves and miniaturization. What are the challenges and opportunities associated with developing nanowave devices?
4. How can waves be harnessed for sustainable energy? Discuss wave energy conversion, ocean thermal energy conversion (OTEC), and advancements in solar and wind energy.
5. Discuss the various applications of waves in medicine and healthcare, including non-invasive therapies, wave-based diagnostics, and the use of sensors. How do these technologies improve medical treatments and diagnostics?

REFERENCES

1. Ando, S. I., Barik, S. K., Feng, Z., Finetti, M., Chaves, A. G., Kanuri, S., ... & Zimmer, F. (2021). Decaying dark matter in dwarf spheroidal galaxies: Prospects for x-ray and gamma-ray telescopes. *Physical Review D*, *104*(2), 023022.
2. Bertozzi, D., Dimitrakopoulos, G., Flich, J., & Sonntag, S. (2015). The fast evolving landscape of on-chip communication: Selected future challenges and research avenues. *Design Automation for Embedded Systems*, *19*, 59–76.
3. Cao, Z., Wen, Q., Wang, X., Yang, Q., & Jiang, F. (2021). An Overview of the Miniaturization and Endurance for Wearable Devices. *Journal on Internet of Things*, *3*(1), 11.
4. Curtis, B. D., & Hurtak, J. J. (2004). Consciousness and quantum information processing: uncovering the foundation for a medicine of light. *The Journal of Alternative & Complementary Medicine*, *10*(1), 27–39.
5. Efobi, U. R., Tanankem, B. V., & Asongu, S. A. (2018). Female economic participation with information and communication technology advancement: Evidence from Sub-Saharan Africa. *South African Journal of Economics*, *86*(2), 231–246.
6. Elayan, H., Amin, O., Shihada, B., Shubair, R. M., & Alouini, M. S. (2019). Terahertz band: The last piece of RF spectrum puzzle for communication systems. *IEEE Open Journal of the Communications Society*, *1*, 1–32.
7. Fox, C. G., & Alder, J. F. (1989). Surface acoustic wave sensors for atmospheric gas monitoring. A review. *Analyst*, *114*(9), 997–1004.
8. Ghanbari, M. M., Tsai, J. M., Nirmalathas, A., Muller, R., & Gambini, S. (2017). An energy-efficient miniaturized intracranial pressure monitoring system. *IEEE Journal of Solid-State Circuits*, *52*(3), 720–734.
9. Ghosh, A., Thomas, T. A., Cudak, M. C., Ratasuk, R., Moorut, P., Vook, F. W., ... & Nie, S. (2014). Millimeter-wave enhanced local area systems: A high-data-rate approach for future wireless networks. *IEEE Journal on Selected Areas in Communications*, *32*(6), 1152–1163.
10. Guerreiro, S. B., Dawson, R. J., Kilsby, C., Lewis, E., & Ford, A. (2018). Future heat-waves, droughts and floods in 571 European cities. *Environmental Research Letters*, *13*(3), 034009.

11. Hackett, L. P., Li, W., Ameen, A., Goddard, L. L., & Liu, G. L. (2018). Plasmonic metal–insulator–metal capped polymer nanopillars for SERS analysis of protein–protein interactions. *The Journal of Physical Chemistry C, 122*(11), 6255–6266.
12. Kellermann, K. I., & Moran, J. M. (2001). The development of high-resolution imaging in radio astronomy. *Annual Review of Astronomy and Astrophysics*, *39*(1), 457–509.
13. Kim, E., Restuccia, F., Yang, J., & Daraio, C. (2015). Solitary wave-based delamination detection in composite plates using a combined granular crystal sensor and actuator. *Smart Materials and Structures*, *24*(12), 125004.
14. Koch, M. (2007). Terahertz communications: A 2020 vision. In *Terahertz Frequency Detection and Identification of Materials and Objects*. Dordrecht: Springer Netherlands. (Vol. 3, pp. 325–338).
15. Liu, L., Larsson, E. G., Yu, W., Popovski, P., Stefanovic, C., & De Carvalho, E. (2018). Sparse signal processing for grant-free massive connectivity: A future paradigm for random access protocols in the Internet of Things. *IEEE Signal Processing Magazine*, *35*(5), 88–99.
16. Loo, C. K., & Mitchell, P. B. (2005). A review of the efficacy of transcranial magnetic stimulation (TMS) treatment for depression, and current and future strategies to optimize efficacy. *Journal of Affective Disorders*, *88*(3), 255–267.
17. Magisetty, R., Shukla, A., & Kandasubramanian, B. (2018). Dielectric, hydrophobic investigation of ABS/NiFe 2 O 4 nanocomposites fabricated by atomized spray assisted and solution casted techniques for miniaturized electronic applications. *Journal of Electronic Materials*, *47*, 5640–5656.
18. Mihók, B., Biró, M., Molnár, Z., Kovács, E., Bölöni, J., Erős, T., ... & Báldi, A. (2017). Biodiversity on the waves of history: Conservation in a changing social and institutional environment in Hungary, a post-soviet EU member state. *Biological Conservation*, *211*, 67–75.
19. Molina, M. O., Sánchez, E., & Gutiérrez, C. (2020). Future heat waves over the Mediterranean from an Euro-CORDEX regional climate model ensemble. *Scientific Reports*, *10*(1), 8801.
20. Neupane, D., Kafle, S., Karki, K. R., Kim, D. H., & Pradhan, P. (2022). Solar and wind energy potential assessment at provincial level in Nepal: Geospatial and economic analysis. *Renewable Energy*, *181*, 278–291.

21. Ng, S. X., Conti, A., Long, G. L., Muller, P., Sayeed, A., Yuan, J., & Hanzo, L. (2020). Guest editorial advances in quantum communications, computing, cryptography, and sensing. *IEEE Journal on Selected Areas in Communications*, *38*(3), 405–412.
22. Perdomo, A., Truncik, C., Tubert-Brohman, I., Rose, G., & Aspuru-Guzik, A. (2008). Construction of model Hamiltonians for adiabatic quantum computation and its application to finding low-energy conformations of lattice protein models. *Physical Review A*, *78*(1), 012320.
23. Pirandola, S., Mancini, S., Lloyd, S., & Braunstein, S. L. (2008). Continuous-variable quantum cryptography using two-way quantum communication. *Nature Physics*, *4*(9), 726–730.
24. Piri, M. M. (2018). The changing landscapes of fintech and regtech: Why the United States should create a federal regulatory sandbox. *Bus. & Fin. L. Rev.*, *2*, 233.
25. Przybylski, A. K., & Weinstein, N. (2013). Can you connect with me now? How the presence of mobile communication technology influences face-to-face conversation quality. *Journal of Social and Personal Relationships*, *30*(3), 237–246.
26. Rajagopalan, K., & Nihous, G. C. (2013). Estimates of global Ocean Thermal Energy Conversion (OTEC) resources using an ocean general circulation model. *Renewable Energy*, *50*, 532–540.
27. Spector, L., Barnum, H., Bernstein, H. J., & Swamy, N. (1999). Quantum computing applications of genetic programming. *Advances in Genetic Programming*, *3*, 135–160.
28. Thongpapanl, N. T. (2012). The changing landscape of technology and innovation management: An updated ranking of journals in the field. *Technovation*, *32*(5), 257–271.
29. Villa, F., Severini, F., Madonini, F., & Zappa, F. (2021). SPADs and SiPMs arrays for long-range high-speed light detection and ranging (LiDAR). *Sensors*, *21*(11), 3839.
30. Wang, L., Lin, M., Tedeschi, E., Engström, J., & Isberg, J. (2020). Improving electric power generation of a standalone wave energy converter via optimal electric load control. *Energy*, *211*, 118945.
31. Yan, L., Fang, X., Wang, X., & Ai, B. (2020). AI-enabled sub-6-GHz and mm-wave hybrid communications: Considerations for use with

future HSR wireless systems. *IEEE Vehicular Technology Magazine*, *15*(3), 59–67.

32. Zacharias, S., Koppe, C., & Mücke, H. G. (2014). Climate change effects on heat waves and future heat wave-associated IHD mortality in Germany. *Climate*, *3*(1), 100–117.
33. Zimmermann, L., Tekin, T., Schroeder, H., Dumon, P., & Bogaerts, W. (2008). How to bring nanophotonics to application-silicon photonics packaging. *IEEE LEOS Newsletter*, *22*(6), 4–14.

INDEX

E

F

G

L

M

N

O

Printed in the United States
by Baker & Taylor Publisher Services